WITHDRAWN

ELECTROSLAG REMELTING
AND
PLASMA ARC MELTING

Report of
The Committee on Electroslag Remelting
and Plasma Arc Melting

NATIONAL MATERIALS ADVISORY BOARD
Commission on Sociotechnical Systems
National Research Council

NATIONAL ACADEMY OF SCIENCES
Washington, D.C.
1976

NOTICE   The project that is the subject of this report was approved by the Governing Board of the National Research Council, whose members are drawn from the Councils of the National Academy of Sciences, the National Academy of Engineering, and the Institute of Medicine. The members of the Committee responsible for the report were chosen for their special competences and with regard for appropriate balance.

This report has been reviewed by a group other than the authors according to procedures approved by a Report Review Committee consisting of members of the National Academy of Sciences, the National Academy of Engineering, and the Institute of Medicine.

This study by the National Materials Advisory Board was conducted under Contract No. MDA-903-74-C-0167 with the Department of Defense.

The National Research Council was established in 1916 by the National Academy of Sciences to associate the broad community of science and technology with the Academy's purposes of furthering knowledge and of advising the federal government. The Council operates in accordance with general policies determined by the Academy by authority of its Congressional charter of 1863, which establishes the Academy as a private, non-profit, self-governing membership corporation. Administered jointly by the National Academy of Sciences, the National Academy of Engineering, and the Institute of Medicine (all three of which operate under the charter of the National Academy of Sciences), the Council is their principal agency for the conduct of their services to the government, the public, and the scientific and engineering communities.

Publication NMAB-324

Library of Congress Catalog Card Number 76-13351
International Standard Book Number 0-309-02505-2

Available from
Printing and Publishing Office, National Academy of Sciences
2101 Constitution Avenue, N.W., Washington, D.C. 20418

Printed in the United States of America

TN
686.5
E4
N37
1976

PREFACE

The National Materials Advisory Board undertook the study reported on in this document at the request of the Department of Defense. Conduct of the study was assigned to the Committee on Electroslag Remelting and Plasma Arc Melting which was charged to assess the position of the U.S. electroslag remelting and plasma arc melting technology vis-a-vis such technology elsewhere. It is germane to inquire why there was such a long lead-time between initial development of the process and its widespread application; why there should have been development in the USSR before in the USA; why the recent growth in the USA should have been so rapid? The purpose of the study is to assess electroslag remelting and plasma arc melting processes and to identify U.S. research and development needs in these areas.

The Committee first met on May 15, 1974 and this report is based on data collected through June 1975.

At its initial meeting, the Committee decided to approach the study by:
1) Reviewing the state of the art of electroslag remelting and plasma arc melting technology
2) Analyzing each melting process to identify specific advantages, process limitations, and comparative cost effectiveness, including energy requirements
3) Recommending specific research and development efforts.

The possible benefits of research and development resulting from such a study are:
1) Improving or developing advanced melting techniques for producing reliable high-strength steels and special alloys by
   a) Lowering processing costs through using molten metal, decreasing alloy loss during processing and the amount of costly alloys used, and scaling-up furnace capacity
   b) Improving mechanical properties through fewer non-metallic inclusions, higher isotropic properties, and improved homogeneity
2) Producing more reliable products and parts at lower cost.

To expedite the study and to provide for proper treatment of the various aspects of the study, the Committee was divided into panels covering those categories.

The Committee particularly wishes to acknowledge the contributions of the late Mr. John Luchok to this study. Mr. Luchok provided considerable input to various sections of the report. His expertise in this field will be sorely missed.

This report is based upon data collected up to June 1975.

NATIONAL MATERIALS ADVISORY BOARD
COMMITTEE ON ELECTROSLAG REMELTING AND PLASMA ARC MELTING

Chairman

MERTON C. FLEMINGS, Ford Professor of Engineering, Department of Metallurgy, Massachusetts Institute of Technology, Cambridge, Massachusetts

Members

GOPAL K. BHAT, Head, Materials and Process Engineering Research, Mellon Institute, Carnegie-Mellon University, Pittsburgh, Pennsylvania

JOSEPH H. KLEIN, Senior Group Leader, Process Metallurgy and Ceramics, Stellite Division, Cabot Corporation, Kokomo, Indiana

*JOHN LUCHOK, Technical Director, NUCOR Steel Division, NUCOR Corporation, Darlington, South Carolina

ALEC MITCHELL, Professor of Metallurgy, Metallurgy Department, University of British Columbia, Vancouver, British Columbia, Canada

RENÉ SCHLATTER, Manager of Research and Development, Latrobe Steel Company, Latrobe, Pennsylvania

RALPH M. SMAILER, Supervisor of Clad and Process Research, Lukens Steel Company, Coatesville, Pennsylvania

Liaison Representatives

O. NORMAN CARLSON, Professor of Metallurgy, Ames Laboratory, Iowa State University, Ames, Iowa

ROBERT HALL, NASA-Lewis Research Center, Cleveland, Ohio

KENNETH LOVE, Program Manager for Casting and Melting, Manufacturing Technology Division, Air Force Materials Laboratory, Wright-Patterson Air Force Base, Ohio

IRVING MACHLIN, Naval Air Systems Command, Washington, D.C.

---

*Deceased.

GEORGE MAYER, Acting Director of Metallurgy and Materials Science Division, U.S. Army Research Office, Durham, North Carolina
RALPH H. NAFZIGER, Research Chemist, Albany Metallurgy Research Center, U.S. Bureau of Mines, Albany, Oregon
FREDERICK A. SCHMIDT, Ames Laboratory, Iowa State University, Ames, Iowa
MICHAEL D. VALENTINE, Naval Air Systems Command, Washington, D.C.
ANDREW VAN ECHO, Energy Research and Development Administration, Division of Reactor Research and Development, Washington, D.C.
EDWARD S. WRIGHT, Associate Director for Operations, Army Materials and Mechanics Research Center, Watertown, Massachusetts

Consultant

HENRY S. RUBENSTEIN, McLean, Virginia

NMAB Staff

BEN A. KORNHAUSER, Staff Engineer

NATIONAL RESEARCH COUNCIL
COMMISSION ON SOCIOTECHNICAL SYSTEMS

NATIONAL MATERIALS ADVISORY BOARD

Chairman:

Dr. Seymour L. Blum
Director
Advanced Program Development
The MITRE Corporation
P.O. Box 208
Bedford, Massachusetts 01730

Past Chairman:

Dr. N. Bruce Hannay
Vice President-Research & Patents
Bell Laboratories
Murray Hill, New Jersey 07974

Members

Dr. James Boyd
Consultant
Materials Associates
Suite 250
600 New Hampshire Ave., N.W.
Washington, D.C. 20037

Dr. Alan G. Chynoweth
Director, Materials Research Lab.
Bell Telephone Laboratories
Murray Hill, New Jersey 07974

Dr. Arthur C. Damask
Professor, Physics Department
Queens College of the City of New York
Flushing, New York 11367

Dr. Herbert I. Fusfeld
Director of Research
Kennecott Copper Corporation
161 E. 42nd Street
New York, New York 10017

Dr. John B. Goodenough
Research Physicist
Lincoln Laboratories
Massachusetts Institute of Technology
P.O. Box 73
Lexington, Massachusetts 02173

Mr. William D. Manly
Group Vice President
Engineered Products
Cabot Corporation
1020 West Park Avenue
Kokomo, Indiana 46901

Dr. Frederick T. Moore
Economist Advisor, Industrial Projects
World Bank
Room F 1010
1818 H Street, N.W.
Washington, D.C. 20431

Mr. Howard K. Nason
President
Monsanto Research Corporation
800 N. Lindbergh Boulevard
St. Louis, Missouri 63166

Dr. Walter S. Owen
Head, Materials Science Dept.
Massachusetts Institute of Technology
Cambridge, Massachusetts 02139

Dr. Harold W. Paxton
Vice President-Research
U.S. Steel Corporation
Pittsburgh, Pennsylvania 15219

Dr. Eli M. Pearce
Professor of Polymer Chemistry and Engineering
Polytechnic Institute of New York
333 Jay Street
New York, New York 11201

Dr. William R. Prindle
8 Pilgrims Path
Sudbury, Massachusetts 01776

Dr. David V. Ragone
Dean, College of Engineering
University of Michigan
Ann Arbor, Michigan 48104

Dr. Rustum Roy
Director
Materials Research Laboratory
Pennsylvania State University
University Park, Pa. 16802

Dr. Raymond L. Smith
President
Michigan Technological University
1400 College Avenue
Houghton, Michigan 49931

Dr. Morris A. Steinberg
Director
Technology Applications
Lockheed Applications
Lockheed Aircraft Corp.
Burbank, California 91520

Dr. Giuliana C. Tesoro
Visiting Professor, Fibers and Polymers Laboratories
Massachusetts Institute of Technology
278 Clinton Avenue
Dobbs Ferry, New York 10522

Dr. John E. Tilton
Associate Professor
Department of Mineral Economy
Pennsylvania State University
University Park, Pa. 16802

Dr. John B. Wachtman, Jr.
Division Chief
Inorganic Materials Division
National Bureau of Standards
Room A359, Materials Building
Washington, D.C. 20234

Dr. Max L. Williams
Dean, School of Engineering
University of Pittsburgh
Pittsburgh, Pennsylvania 15213

NMAB Staff

C. W. Spencer, Executive Director
J. V. Hearn, Executive Secretary

CONTENTS

I. SUMMARY OF CONCLUSIONS AND RECOMMENDATIONS ... 1
   A. Conclusions 1
   B. Recommendations 4

II. INTRODUCTION 7
   A. Processes 7
   B. Soviet-U.S. — ESR/VAR Technology in
      Perspective 11
   C. Historical Review 12

III. STATUS ASSESSMENT OF ELECTROSLAG REMELTING ... 15
   A. The Western Hemisphere 15
   B. Western Europe 16
   C. Japan 17
   D. The Soviet Union 18
   E. The United States 21

IV. GENERAL PROCESS DESCRIPTION 23
   A. Principles of Electroslag Remelting Process ... 23
   B. Basic Process Parameters 24
   C. Slags and Process Chemistry 25
   D. Basic Furnace Designs 28
   E. Energy Requirements 29
   F. Economic Aspects 31
   G. Shapes 32
   H. Environmental Considerations 33

V.	ELECTROSLAG REMELTING TECHNOLOGIES	35
	A. Electroslag Ingot Technology for Superalloys	35
	B. Electroslag Ingot Technology for Carbon and Low-Alloy Steels	48
	C. Electroslag Ingot Technology for Stainless and High-Alloy Steels	63
	D. Electroslag Technology for Other Metals and Alloys	73
	E. Electroslag Methods for the Manufacture of Large Ingots	88
VI.	NOVEL APPLICATIONS OF ELECTROSLAG TECHNOLOGY	115
	A. Introduction	115
	B. Novel ESR Process Adaptations	116
VII.	ELECTROSLAG EQUIPMENT TECHNOLOGY	133
	A. Current Equipment Technology	133
	B. Trends and Forecast of Developments	142
	C. Conclusions and Recommendations	145
VIII.	PLASMA ARC MELTING TECHNOLOGY	149
	A. Introduction	149
	B. Plasma Arc Melting	149
	C. Plasma Arc Melting Furnaces	153
	D. Plasma Melting Process Economics	163
	E. Conclusions and Recommendations	164
IX.	CRITICAL EVALUATION OF ELECTROSLAG AND PLASMA ARC MELTING	167
	A. Electroslag Melting	167
	B. Plasma Melting	171

APPENDIXES

A.	GLOSSARY	175
B.	ELECTROSLAG FURNACE INSTALLATIONS	179
C.	REFERENCES AND BIBLIOGRAPHY	191

TABLES

Table

1. General Characteristics of Soviet ESR Furnaces 20
2. Installed U.S. Electroslag Remelt Capacity Estimated for 1975 21
3. Projected Capacity Increase and Market Application for ESR Materials in the United States for the Year 1980 22
4. Effect of ESR Power Mode on Power Consumption 30
5. Typical Superalloys Melted by ESR 35
6. Some Important Equipment and Process Variables in Electroslag Remelting of Superalloys 36
7. Results of Chemical Analysis for Selected Elements Al, Si and O Compared to Electrode Values of 0.15, 0.45 and 0.002, Respectively 40
8. Mechanical Properties of HASTELLOY X Bar 42
9. Comparison of AMS 5596C Mechanical Property Test Results for ESR and VAR Flat Products of Solution Treated and Annealed HAYNES Alloy 718 44
10. Inclusion Count Rating of Several Research Project Steels, Lectrefine® Processed 51
11. Chemical Analysis of Some Research Project Steels 53
12. Chemical Analysis in Weight Percent of Lectrefine® A533B (Heat R0048) 55
13. Inclusion Count – JK Ratings per ASTM E-45 Plate III (Heat R0048) 55
14. Room Temperature Tensile Properties of the Quarterline of 6-inch Gauge Lectrefine® A533B (Heat R0048) 55
15. Tensile Data for Electroslag Remelted A533B (Heat R0048) . . . 56
16. Summary of Charpy V-Notch Impact Properties of the Quarterline of 6-inch Gauge Lectrefine® A533B (Heat R0048) 56
17. Fatigue Data for Transverse Quarterline of 6-inch Gauge Lectrefine® A533B (Heat R0048, Low Cyclic Rate) 57

Table

18	Fatigue Data for Transverse Quarterline of 6-inch Gauge Lectrefine® A533B (Heat R0048, High Cyclic Rate)	58
19	Fatigue Crack Growth Rate ($\frac{da}{dN}$) for Electroslag Remelted Lectrefine® A533B (Heat R0048)	59
20	Fracture Toughness Data for Electroslag Remelted Lectrefine® A533B (Heat R0048)	59
21	Typical Flux Compositions for Remelting Alloy Tool Steels	66
22	Average Impact Strengths of Plate Fabricated from Electroslag Melted and Vacuum-Arc Melted Titanium and Zirconium Ingots	83
23	Average Values of Melting Rates and Energy Requirements for Melting Reactive, Refractory, and Heavy Metals by Various Processes	85
24	Mechanical Properties of ESR Cast Hollow Ingots and Wrought Rings of Low-Alloy Ultra-High-Strength Steels	124
25	Mechanical Properties of ESR Cast Hollow Ingots and Wrought Rings of Heat-Resistant Alloys and Superalloys	125
26	Direct Electroslag Melting	131
27	Comparison of Different ESR Power Modes	138
28	Comparison of Slab Ingot Production ESR Furnace Characteristics	139
29	Properties of Gases Used in Plasma Applications	151
30	Alloy Recovery in Plasma Induction Melted Metal (Daido Steel Data)	157
31	World Electroslag Furnace Installations	180

ILLUSTRATIONS

Figure

1	Schematic of the Electroslag Remelting Process	8
2	Schematics of Basic ESR Furnace Designs	28

Figure

3	Comparison of Costs of Forged Round Billets from ESR and Normal Ingots Produced with the Same Forging Reduction, and a Lower Reduction Ratio for ESR Ingots	32
4	Distribution of Oxide Inclusion Ratings for 0.040- to 0.080-inch in Sheet of HASTELLOY X	39
5	Gleeble Hot Ductility of Udimet 700	41
6	Power Requirements and Melt Rates versus Ingot Size for Electroslag Remelting of Alloy Tool Steels – Effect of Power Mode on Average Power Requirements	65
7	Hot Tensile Properties of Air-Melted and ESR M-2 High-Speed Steel	69
8	ESR and VAR Molybdenum Ingots	75
9	Comparative Macrostructures of ESR and VAR Molybdenum Ingots	77
10	Ductile-to-Brittle Transition Temperature of 0.15 cm Sheet Rolled from Electroslag and Arc-Melted Molybdenum Ingots	79
11	Tensile Properties and Hardness of 0.16-cm-Thick Titanium Sheet Rolled from Vacuum-Arc Melted Material and Electroslag Melted Material	81
12	Thermal Profile for an 80-Inches-Diameter Ingot Melted in a Collar Mold at 4,400 lb/hr	90
13	Instantaneous Melting Rate as a Function of Time for Two Diameters of ESR Ingots	91
14	Electrode Geometry Configurations and Power Input That Have Been Used for Melting Large ESR Ingots	93
15	Multi-Electrode Reduced Section of Mold Configuration	99
16	Heat Transfer Regime at the Copper/Water Interface in the Process Region	102
17	Central Zone Remelting Process	109
18	Manufacturing Concept of Small-Cross-Section ESR Ingots from Large-Cross-Section Electrode	117
19	Manufacturing Concept of Multiple ESR Ingots from a Large-Cross-Section Electrode	118

Figure

20	Schematic Illustration of Roll Manufacture Via ESR	119
21	ESR Melting of Composite Metal Components	119
22	ESR Metal Overlay or Surface Repair Arrangement	120
23	Bifilar ESR Furnace Setup for Making Square and Rectangular Ingots	121
24	Schematic Representation of ESR Hollow Ingot Manufacturing Technique through Melt Extrusion	123
25	Schematic Representation of ESR Hollow Ingot Production Setup Using Funnel Mold and Polyelectrodes	126
26	Essential Features of an ESR Setup for Manufacturing Pressure Vessels with Integral Dome	128
27	Diagram of Electroslag Melting of Tube Billets by the YOZO Method	129
28	Schematic Arrangement of Electroslag System for Scrap Consolidation	130
29	Principal Ways of Supplying Power for the Electroslag Process	134
30	Three-Station ESR Furnace for 20- to 40-Inch Diameter Ingot	136
31	Lukens Steel ESR Slab Furnace of CONSARC Design	137
32	Schematic Diagram of the CESPM Process	140
33	Continuous ESM Furnace and Caster	141
34	Schematic Diagram of a Three-Phase CESPM Unit for Large Ingots	142
35	Direct Current Transferred-Arc Plasma Torch Schematic	150
36	Plasma Temperature as a Function of Gas Energy Content at Atmospheric Pressure	152
37	Cross Section of the Plasma Arc Furnace	154
38	Plasma Induction Furnace	157
39	Technological Scheme of Plasma Arc Remelting – Soviet Design	158

ABSTRACT

An assessment is made of the electroslag remelting (ESR) and plasma arc melting (PAM) technologies used in the United States in manufacturing a variety of materials and sizes of ingots. Significant metal quality improvements in surface condition, mechanical properties, solidification, structure cleanliness, and yield over air-melted material have accelerated ESR usage, particularly abroad. PAM has potential in recycling scrap, producing complex alloys, large monocrystals of refractory metals and their alloys, and independent control of heat input and metal feed. ESR has found significant applications in producing premium quality high alloy and specialty alloy steels, providing features such as desulfurization, retention of volatile alloying elements, and the capability to produce shaped ingots and castings. Many advantages and disadvantages are listed along with research and development to improve the technology and usefulness.

I. SUMMARY OF CONCLUSIONS AND RECOMMENDATIONS

A. <u>Conclusions</u>

1. The industrial status and potential of the electroslag remelting (ESR) process are summarized below.
    a. During the past 15 years, U.S. production of electroslag remelted ingots has increased rapidly from less than 10,000 tons per year to approximately 126,000 tons per year, an amount representing about 25 percent of current world ESR production capacity. This growth was stimulated by the advantageous technology of the process and the demand for the high-quality materials that could be produced.
    b. The materials currently produced using the ESR process and their estimated percentage of 1975 installed capacity are as follows: carbon and low-alloy steels, 22 percent; tool and die steels, 27 percent; stainless steels, 20 percent; nickel- and cobalt-base superalloys, 31 percent.
    c. The demand for ESR materials is projected to reach 500,000 tons per year by 1980. The most substantial growth is expected in the demand for carbon and low-alloy ESR steels--from an estimated 28,000 tons* per year in 1975 to 255,000 tons per year in 1980. This projection of a ninefold increase during the next five years is based on anticipated cost-effective use of ESR low-alloy steels for applications such as rotor steels, roll alloys, armor plate, ship plate, gun tubes, gears, bearings, and high-performance aircraft structural members and a demand for increased performance from traditional low-alloy steels that conserve strategic alloy elements.
    d. Meeting the projected demand for ESR materials in 1980 will require a substantial capital investment and even though the specific projection is speculative, it is certain that the current rate of growth in production capacity will be exceeded.

---

*The ton used herein is the short ton of 2,000 pounds.

e. The rapid projected growth of ESR capacity is expected to have only a small impact on overall energy needs and the environment.

2. The quality of ESR ingots is significantly better than that of air-melted and conventionally cast ingots. The greater uniformity and more homogeneous structures of ESR ingots result in improvements in mechanical properties (e.g., better fracture toughness and ductility and less directionality), surface condition, solidification structure, cleanliness, and yield.

3. The ESR process offers a number of advantages over other consumable electrode remelting processes including the capability to:
   a. Decrease sulfur content.
   b. Achieve faster, more directional solidification with the result that the structure improves (e.g., inclusions are finer and more evenly distributed).
   c. Make slabs, hollows, and other shapes.
   d. Produce structures with good homogeneity (because of the high melting stability).
   e. Improve surface quality and increase yield.

   The potential for making large-sized ingots at reasonable production rates also exists.

4. The disadvantages of current ESR technology are that:
   a. Control of hydrogen and other trace impurities is poor.
   b. Control of reactive alloy elements is limited.
   c. Power consumption is higher than in vacuum arc remelting (VAR).

5. Further research and development are expected to result in significant advances in ESR processes by improving understanding of thermal and solidification behavior, metal refinement, and scaling the process to larger sizes.

6. The basic ESR process for production of superalloys generally is well understood and appears to offer significant advantages in the production of some difficult-to-fabricate alloys (e.g., advanced gamma-prime-strengthened superalloys used extensively in aircraft gas turbines). Theoretical understanding of ESR process details, especially solidification structure control, requires improvement, and methods for reliably controlling reactive elements and removing harmful trace elements are needed.

7. The current importance of ESR superalloys is in their use in static components of aircraft gas turbines. Wider application is restricted

by materials specifications requiring use of vacuum melted materials, particularly in rotating components where expensive qualification engine testing is required. Qualification of ESR materials depends on the demonstration of economic and technical incentives sufficient to justify engine testing.

8. Use of the ESR process to produce carbon and low-alloy steels is relatively advanced. The largest size slab cross section currently produced in the United States is 30 by 60 inches; however, it is expected that 30 by 80 inch slabs will be produced in the near future. Capital costs per ton are high and productivity is lower than that of conventional air melting and casting. Solidification shrinkage is causing difficulties in the hot topping of large ingots, and hydrogen control is a serious problem.

9. ESR has been used considerably in the production of premium-quality high-alloy steels and specialty alloys. The feasibility of using ESR rather than conventional air melting for many high-alloy steels depends upon whether the remelting costs can be offset by major metallurgical, technological, or yield benefits. The ESR process is, however, an alternative to the VAR process and provides such advantages as desulfurization, retention of volatile alloying elements, and a wide technological flexibility with the capability to produce shaped ingots and castings.

10. No commercial electroslag operations for melting reactive, refractory metals or alloys are known to exist and none are anticipated in the near future.

11. Current ESR technology permits the manufacture of 65-ton round-section ingots (5 feet in diameter by 16-2/3 feet in length) and ingots of up to 110 tons have been produced on a developmental basis. The soundness and mechanical properties achievable in large ESR ingots are clearly superior to those of ingots produced conventionally. To scale the process up to much larger ingots, furnace design and chemical and solidification control problems must be solved.

12. Although critical evaluation has not been made yet, it appears that five processes are available, or are potentially available, for making large ingots: conventional air melting and casting, conventional ESR, electroslag (ES) casting, bifilar welding, and central zone remelting (CZR).

13. Hollow ingots and more complex castings, such as crankshafts and valves, can be made using the ESR process. In the United States, efforts to date have concentrated primarily on the production of hollow ingots for ring pre-forms and tube blanks, and little process or equipment development effort has been devoted to shape casting.

14. Ingot production equipment technology is highly developed and versatile, reliable furnaces are now in operation. The design of improved furnaces requires that a critical study be made to design molds with improved life, assess full length vs. collar molds, determine electrode change effects on solidification structure, and increase power efficiency.

15. Evidence indicates that as-cast ESR materials offer better mechanical properties than conventional castings and wrought products.

16. Despite years of research and development, the technological and equipment difficulties associated with the plasma arc melting and remelting processes thus far, have prevented their commercial application in the United States and clear-cut economic or technical advantages have not been demonstrated yet. Potential applications are in the recycling of high-alloy scrap and the production of ingots of complex alloys and large monocrystals of refractory metals and their alloys. Plasma arc remelting furnaces, however, do offer potential advantages over ESR furnaces in that heat input and metal feed rate can be controlled independently (although structure refinement by this means would require reduced melting rate) and atmosphere control is flexible.

B. Recommendations

It is recommended that research and/or development efforts be undertaken to:

1. Characterize the properties of wrought and cast ESR material in relation to the equivalent conventional material, with particular emphasis on those properties strongly influenced by ESR. Relate property improvements to process and structural variables.

2. Make a critical systems evaluation of the five available, or potentially available, processes for making large ingots: conventional air melt and cast, conventional ESR, electroslag casting, bifilar welding, and central zone remelting.

3. Study the physical chemistry of ESR slags and the kinetics of removing impurities and introducing beneficial trace elements. Study electrochemical reactions in detail. Develop low CaF_2 slags and techniques to recycle spent slags to counteract the growing shortage and high cost of high-grade CaF_2.

4. Model the thermophysical behavior during melting and solidification in the ESR system to permit improvement in quality and economy of existing ingots and scale-up to larger sizes.

5. Continue current work on ESR hollow ingots, and examine the economic feasibility of extending this work to other shapes.

6. Improve solidification control to permit large cross-section ingots with a high degree of structural refinement and consistency to be produced in complex structure-sensitive alloys. One problem to be overcome is the hot-topping of very large ingots.

7. Develop improved model technology for ESR. Evaluate other furnace design factors including the benefits of full vs. collar molds, the effect of electrode change on solidification structure, and feasibility of developing a furnace with molten feed. Consideration also should be given to increasing power efficiency.

8. Develop a practical process for casting several small ingots simultaneously from one large electrode.

9. Develop reliable techniques for reducing hydrogen and other impurities to acceptable low levels.

10. Examine the feasibility of employing plasma arc remelting in recycling specialty and high-alloy scrap and producing ingots of complex alloys and large monocrystals.

11. Develop plasma torches with improved characteristics and reliability for use as auxiliary heating devices in metallurgical systems such as induction melting, degassing, and refining.

II. INTRODUCTION

A. Processes

Remelting processes have been used in the special-alloy production sequence for over 30 years. Until recently, the vacuum arc remelting (VAR) process was the most widely used in the United States, but during the past 10 years use of the electroslag remelting process (ESR) has increased rapidly. Plasma arc melting and remelting processes occupy a somewhat different position because they are already accepted as metallurgical methods in the areas of technology not necessarily requiring a consumable electrode technique. Probably, the major application of the plasma arc processes lies in areas that are not simply alternatives to the conventional VAR and ESR routes.

1. Electroslag Remelting

The concept of the electroslag remelting process* is extremely simple. A solid electrode is made one pole of a high-current source and a water-cooled baseplate or crucible is the other pole. The connection between the two poles occurs through a conducting slag contained in the water-cooled mold. The slag acts as ohmic resistance and the Joule heating produced in it melts the electrode tip. The slag compositions used are vital to the correct operation of the process; generally, they fall within the system $CaF_2 + CaO + Al_2O_3$ and fulfill the basic conditions (e.g., electrical and thermal conductivity, high-temperature stability, chemical effects, and phase behavior) imposed by the process requirements. The electrode is melted progressively into the crucible through the slag, and the liquid metal also solidifies in a progressive fashion (see Figure 1 for a schematic of the process). It resembles the VAR process in that:
1) Refining occurs in a small high-temperature zone
2) The electrode is melted progressively
3) Incremental solidification occurs in a water-cooled mold.

---

* A glossary of many ESR terms is presented as Appendix A.

It differs primarily in that:
1) Processing occurs at atmospheric pressure
2) The slag and metal are in contact
3) Resistance rather than arc heating is used
4) Ingot solidification occurs inside a slag-skin rather than by limited direct contact with the crucible wall in a low-pressure environment.

The similarity of the ESR solidification mechanism to that of the VAR made it a competitor of the VAR technique and, at an early developmental stage, aircraft-quality alloys were made satisfactorily using the Hopkins ESR technique. However, since ESR melting occurs at atmospheric pressure, no interstitials can be removed by evaporation. Processing titanium, zirconium, uranium, etc., is much more difficult using the ESR rather than the VAR process. The VAR process, therefore, benefitted from reactive metal processing technology and developed more rapidly in the United States. In the U.S.S.R., however, use of the VAR process was not prevalent and the Soviets more fully investigated the potential benefits of the ESR method, resulting in a discrepancy in technology between the two countries in the late 1950s and 1960s.

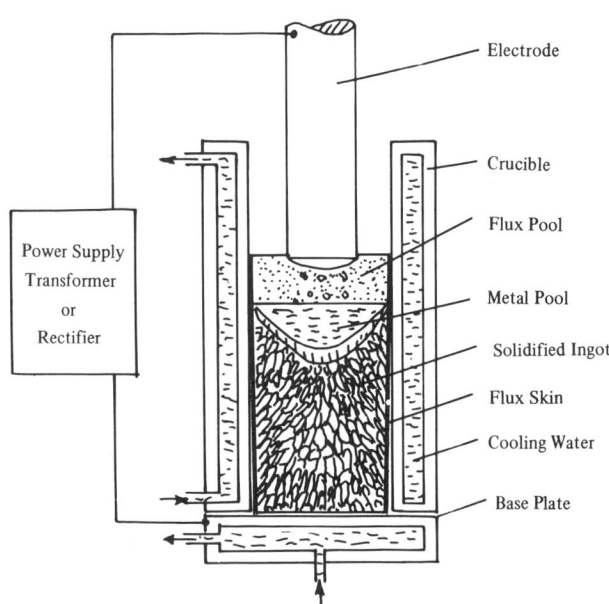

FIGURE 1 Schematic of the Electroslag Remelting Process (from Schlatter, 1972).

A number of reasons can be found for the recent upsurge of interest in the ESR process in the United States including the fact that the slag/metal contact provides a significant increase in the potential range and extent of refining reactions, classically with respect to desulfurization. While the method provides a good finishing route to low-sulfur alloys from conventional air-melt electrodes, the advent of new, lower cost desulfurization processes occurring in the ladle (e.g., as calcium-argon blowing and ladle refining) makes it doubtful that any new ESR capacity would be installed primarily for desulfurization purposes. The ESR process does, however, produce ingots of significantly higher quality (in terms of surface, cleanliness, and structure*) than air-melt processes and the yield is higher. It is for these reasons that the ESR process has been used widely, particularly in the Soviet Union, as an improving step for poor-quality open-hearth steels. The improvements are not as marked over good U.S. arc-furnace practice. With respect to VAR, the major goal of ESR application has been to reproduce the properties of the VAR material at a higher yield and production rate. Thus, when evaluating the role of ESR, it is important to consider it in the correct context, and any comparisons between ESR material and others must recognize the appropriate alternatives.

The Joule heating of the ESR process permits utilization of line-frequency power while all industrial VAR equipment requires direct current, and initially this seemed to give ESR an economic advantage over VAR. To obtain the same ingot quality, however, the ESR process appears to require the same level of power stability and control as the VAR process and in specialty alloys at least, this requirement largely eliminates the cost difference between the ESR and VAR power supplies. At this stage of development, a further potential advantage accrues from the inherent stability of ESR Joule heating in that large electrode sections may be melted with much less difficulty than in the VAR process. The economic advantage of this stability is difficult to assess since it is only one of the advantages ESR offers in large-section ingots.

The ESR process can be made to operate at higher melting rates than the VAR process because of its more stable heating mode and higher heat dissipation rate in the process region. The specific energy efficiency (lb/kWhr) of ESR is less than that of VAR by about 20 percent when good practices are used in both processes. While the higher ESR amortization rate (arising mainly from higher melting rates) cannot offset this disadvantage, it is clear that process development potentially can increase ESR energy efficiency and that ESR offers higher productivity than VAR. For this reason as well as technical ones, new ESR installations in the western world have outnumbered VAR facilities by a large factor during the past 10 years.

The ESR process must operate at pressures greater than the vapor pressure of the slag; in a chemical sense this means at pressures insignificantly

---

*The general term "structure" as used in this report encompasses those features of the ingot that relate solidification to mechanical properties — i.e., macro- and micro-segregation, macro- and micro-porosity, dendrite spacing, and nonmetallic inclusion morphology and distribution.

different from one atmosphere. A range of 1/3 to 10 atmospheres has been reported in the literature, but virtually all processes operate at one atmosphere and in air. Although the prime chemical advantage offered by the VAR process has been lost, the restrictions that largely prevented some very useful modifications of the VAR technique have been removed in that:
1) Other than circular ingot cross sections can be made
2) Electrode changing and melt turn-around are easy as is visual control
3) Movement between ingot and crucible is relative
4) Chemical modifications can be made by additions during processing.

These modifications were useful in ESR.

2. <u>Plasma Arc Melting and Remelting</u>

Small plasma torches are in wide use as cutting and welding devices. Large torches (up to 20 megawatts [mw]) are in industrial use as gas heating devices, primarily in chemical manufacture. Torches have been used as graphite-electrode substitutes in refractory-lined furnaces of a design resembling the conventional electric-arc furnace. Such furnaces, designed and constructed by Daido Steel, are used in production in Japan. Plasma also has been used as an ancilliary heating source in induction furnaces operating with slag additions. Several development-scale devices in which one or more torches act as a heat source for a consumable melting technique have been built. Given this wide range of applications, it is difficult to find common themes of development; however, the most obvious is torch development. The heating efficiency of the nontransferred direct-current plasma torch, with respect to the melting of an electrically neutral target workpiece, is very small and acceptable only for some cutting and welding purposes. If the plasma is treated essentially as an atmospheric-pressure electron beam and heating is caused by a transferred-arc plasma torch in which the target is one pole of the heating circuit, the efficiency becomes much greater. This mode is used most commonly, with either direct or alternating current, in small devices. Very large direct-current torches have been made, but the larger systems presently appear to rely on alternating current with a three-torch, three-phase-delta configuration. As the problems of dealing efficiently with cathode life and anode power dissipation have been investigated and ameliorated, a corresponding increase in both torch reliability and efficiency has been observed. At this time, the plasma method appears a feasible alternative high-temperature heating method for a wide range of applications. Plasma arc remelting should be compared primarily with electron-beam heating rather than with ESR or VAR because the heating rate is independent of the metal feed rate and the heat source can be directed externally to any process area. These qualifications are potentially extremely useful, for example, in the manufacture of segregation-sensivite reactive metal ingots.

In other than remelting applications, plasma is essentially a convenient heat source that does not dictate any specific atmospheric condition. Torches may be used as ancilliary heating in a vacuum environment (e.g., VIM) if the torch gas flow and the system pumping are balanced properly. Plasma torches also may be used as a heat source in skull-melting under vacuum for scrap recovery; in such applications, they offer a potentially simpler method than either the consumable or nonconsumable arc methods.

B. Soviet-U.S. — ESR/VAR Technology in Perspective

One source of recurrent problems in understanding how ESR is used lies in the process' flexibility and extremely broad tolerance for technical level of operation. For example, ESR in aircraft-quality superalloy ingots result when melt-rates are minized to avoid defects, extreme care is taken in process control, and, in many instances, vacuum-induction melted electrodes are used. In this application, the process is highly specialized, small scale, expensive, and directly competitive with VAR. At the other end of the application spectrum, the ESR furnace makes 110-ton ingots (7.5 feet diameter [∅] by 16.7 feet long) of low-alloy steel; in this case, the electrodes are air-melted and the ingots produced have significant segregation and large inclusions by aircraft standards but are improved significantly over those produced by conventional practice. The ESR furnace application has no actual or potential counterpart in VAR and it appears to be a large-scale steelmaking operation that permits wide control variations and requires acute attention to cost details that are of less importance in smaller applications. Although ESR production of both aircraft-quality superalloys and low-alloy steel involves the same basic process, the applications, development directions, and potential uses are so totally different that this report is structured to underline these differences and focus on use areas.

Another problem arises when assessing the past sequence of ESR development because of some rather extreme attitudes that were adopted with respect to the process—that is, both the economic and technical advantages of ESR have been exaggerated and biased comparisons of highly developed VAR techniques and the results from primitive ESR equipment have been made and vice versa. Credibility has been strained further by incomplete technical assessments of Soviet achievements in ESR and by unrealistically optimistic expectations of the pace and success of ESR development in the West. This study hopes to present an unbiased and unambiguous picture of the present ESR status and to correct most of the above misapprehensions. The conclusions, recommendations, and evaluation outline the present and potential merits of the process in comparison with realistic alternatives. They also discuss the proven drawbacks of the technique and areas where its application is likely to prove unsuccessful.

It is also necessary that use of the two processes (ESR and plasma) be put in proper perspective. If ESR fulfills its most optimistic proponents' projections, it will constitute a process step for only 0.25 percent of the U.S. steel production. Its annual energy usage will be minute in the overall context; its environmental impact, at worst, a matter of operator protection; and its total capital absorption less than one modest mini-mill. On the other hand, the material processed could constitute almost all of the nation's capacity in high-quality alloys, that are necessary for the maintenance of energy production, the transportation industry and military capability.

C. Historical Review

A resistance-heated slag or flux had been used in welding for several decades. In the United States, R. K. Hopkins started to melt and refine alloys under slag in 1935 and patented the Kellogg Electric Ingot Process (Hopkins, 1940). Hence, he is considered the inventor of the "electroflux" melting process and his many patents in this field are well recognized. The original process consisted of melting a tubular electrode formed continuously from strip and filling it with granulated alloy additions in proportions that would produce ingots with the desired alloy composition (Hopkins, 1948). The presence of unmelted fragments in the ingot forced Hopkins to switch to solid consumable electrodes in early 1950. The large increase in vacuum arc remelting capacity during the 1950s and 1960s overshadowed the moderate output of the only Hopkins melting installation operated by the former Firth-Sterling Company from 1959 to 1968 (McKeen et al., 1962); however, studies at the Mellon Institute under an Air Force contract (1971) stimulated renewed interest in consumable electrode remelting under a molten flux blanket. Accelerated development of production facilities and remelting technology was stimulated further when Firth-Sterling decided to license other steelmakers in 1964.

Development work in the Soviet Union started in about 1954 when evaluations of electroslag welding at the E. O. Paton Welding Institute in Kiev revealed that alloys with excellent technological properties could be produced. Operation of production furnaces began in about 1958 and a strong effort was made to increase the use of the refining method. In the early 1960s the Soviets began to publish their results, indicating an excellent capability of the process for the production of high-quality alloys in competition with the vacuum-arc process (Leybenzon and Tregubenko, 1963). These reports generated interest in Western countries, and a number of development projects were started.

Large-scale evaluation and development of electroslag remelting began in about 1963 in England at Firth-Brown, at the English Steel Corporation and at the British Iron and Steel Research Association. Significant research work that culminated in the development of uniquely designed equipment and advanced melting technology (Holzgruber and Ploeckinger, 1968) also was carried out at

Boehler Brothers in Kapfenberg, Austria. Several production installations were started in other European countries, and the growth of production capacity has been continuing at a rapid rate. A major effort to further develop electroslag refining has been expended in Japan in recent years. Several large industrial installations are now in operation, and considerable progress in technology, production capability, and application has been reported (Kusamichi and Fukuhara, 1966). During the past 10 years, equipment and process technology have been developed to high standards in the United States and ESR is now a well accepted practice for a variety of specialty steels and alloys in various product forms.

III. STATUS ASSESSMENT OF ELECTROSLAG REMELTING

This section covers the current status of worldwide ESR production capability and planning. The information presented covers the Western Hemisphere as a whole (United States, Canada, and South America), Western Europe, Japan, the Soviet Union, and the United States. Included are brief descriptions of current national interest, research efforts, and production philosophy. A list of all known ESR production installations is presented in appendix B.

A. The Western Hemisphere

1. National Interest and Research

Western Hemisphere interest in the ESR process has focused primarily on the specialized alloy industry: stainless and high-alloy steels and nickel- and cobalt-base superalloys. This interest developed slowly for two reasons: First, in the specialized alloy industry, ESR competes with the well developed vacuum arc remelting process. Second, in the cost- and production-rate-sensitive carbon and low-alloy steel market, ESR had to prove itself cost and quality competitive with an advanced electric furnace and ingot casting technology.

Western Hemisphere interest in ESR has increased rapidly during the past 10 years with the installation of approximately 21 production and 8 research furnaces and the conversion of at least 6 VAR furnaces for direct current ESR. At least 5 new production furnaces are expected to be installed in 1975. During the past 5 years, interest in the electroslag remelting of carbon and low-alloy steels to produce very large ingots, high aspect ratio slab ingots for plate products, and hollow ingots has increased and is expected to continue to grow with the production of carbon and low-alloy steels commanding a substantial fraction of total ESR capacity.

2. Production Trends

ESR technology was influenced significantly by vacuum arc remelting technology. VAR may be characterized by the use of full-length static molds, single-phase AC power supplies with continuously variable control by saturable reactors, and high area fill ratios (defined as exceeding an electrode-to-ingot cross-sectional area ratio of 0.65).

Full-length static molds produce higher quality ingot surfaces than collar molds of either ingot withdrawal or moving mold design and they have the capability of producing high aspect ratio slab ingots. They do, however, represent a substantial capital investment. Continuously variable, or stepless, power supplies offer advantages for the production of segregation-prone alloys in which thermal stability (a major advantage of the ESR process) is of critical importance. The tendency is to use high area fill ratios and static molds where runouts are not a problem. High fill ratios provide for a minimum of specific power consumption and when large static molds are used, the specific power consumption may be as low as 700-800 kWhr per ton, compared to specific energy consumptions of 1000-1400 kWhr per ton for large moving collar molds.

The rapid rate of increase in ESR capacity is expected to continue during the next 10 to 15 years. Current production of special alloys should expand with increasing emphasis placed on high aspect ratio slabs and hollows for rolling and forging preforms. Future developments also may include the development of processing techniques for the simultaneous casting of a number of small specialty ingots from one large electrode.

Activity in the electroslag remelting of carbon and low-alloy steels also should increase considerably and should include the production of large tonnages of low-alloy steel plate products from slab ingots, the production of low-alloy steel hollows, the cladding and production of rolls, and the production of very large ingots for forging preforms.

B. Western Europe

1. National Interest and Research

With the possible exception of France, which has a modestly sized but highly developed VAR capability for the production of super alloys and other specialty steels, Western European nations that possess VAR steel and specialty alloy making capability have not allowed VAR to become as dominant a processing technique as has the United States. Perhaps the need for specialty melting arose later for the European countries and coincided with the sudden upsurge of ESR interest and development in the 1960s.

In recent years, use of VAR has continued in Europe but ESR processing increasingly has dominated the scene as an extension of specialty steel processing. Ingot structure control through progressive solidification was applied to produce a wider range of alloys and larger ingots of low-alloy steels.

Some of the earliest and finest ESR research and development work was conducted in the United Kingdom beginning in the early 1960s, and such fundamental process research is continuing. In addition, the largest ESR furnace in the world is operational in Germany; it can produce ingots up to 92 inches in diameter, but ingots of 60 inches in diameter usually are made. While it is premature to assume that ESR ingots of 120 inches diameter, weighing about 300 tons, will be made in the near future using conventional ESR processing, substantial research is under way to explore that possibility. Work also is being conducted on three-phase melting of high-aspect-ratio alloy steel ingots, and Birlec Ltd. currently is building a 50-ton, three-phase, three-electrode, electrode change ESR furnace for British Steel Corporation.

Remelting of used rolls is routine in Western Europe while as-cast ESR rolls and cast roll sleeves are being developed. The remelting of high nitrogen steels (up to one percent) in a pressurized ESR furnace is being studied.

2. Production Trends

The ESR process is accepted widely in Western Europe and is used to produce alloy products whenever overall economic justification is sound. In general, European ESR products do not command premium market prices unless a technical advantage for the end user is clearly evident and quantifiable.

C. Japan

1. National Interest and Research

The Japanese metals and steel industries have a broad interest in the ESR process. (Japanese production capability is included in appendix B). Many pilot or pre-production units are being operated to permit a full evaluation of the technical and economic aspects of ESR processing for a wide range of alloy applications. In addition, the Japanese are operating a number of research and development units (not listed in appendix B).

Large forging-grade steel ingots (up to 40-ton slabs) currently are produced as are relatively small ingots of special steel (e.g., tool, bearing, and high-alloy steel), nickel-base alloys, and alloys with special electrical and magnetic properties. Existing ingot size capability is adequate for production of most of these specialty grades although larger ingot sizes and newer and more efficient ESR furnaces would improve the process economics substantially.

The Japanese also are interested in applying ESR to tonnage alloys such as alloy steel plate for ships, reactor vessels, and other critical end product applications. Aware of current limitations of conventionally produced steel for large forging applications, they are attempting to expand their ESR ingot-size capability beyond the present 40-ton limitation.

Nippon Steel recently started operating a 40-ton ESR slab furnace of Soviet bifilar design with a moving mold. While this is a production-size furnace, it also will serve as a pilot unit for the design of even larger furnaces. The Japanese also are studying the possibility of low-cost ESR production of stainless steel and heat-resisting alloys (particularly for plate and sheet products); however, the energy crisis has slowed this effort to exploit the ESR process on a larger scale.

2. Production Trends

The ESR process is accepted widely and is used to produce alloy products of various ingot sizes and shapes whenever the overall economic or technical justification is sound. The Japanese are evaluating the ESR process on a broad basis with ingot sizes that can be scaled reasonably to larger "production sizes." With the exception of certain specialty steels that must be limited to small ingot sizes, the high-production phase of ESR processing in Japan is yet to come.

D. The Soviet Union

1. National Interest and Research

The USSR leads the world in the widespread application of ESR in the manufacture of specialty steels and the more conventional steel grades commonly used for critical applications. Additionally, the Soviets have led in the development and industrial application of some of the more novel ESR techniques (e.g., hollow ingots and shape castings). Since its commercial introduction in the late 1950s, ESR processing has made a considerable contribution to overall Soviet steelmaking capability and material quality and the Soviets are expected to expand the use of ESR technology in the metal industry.

In general, Soviet ESR research efforts are aimed at improving metal quality and achieving higher processing efficiencies. For example, the Soviets are studying the use of molten metal as a substitute for cast and wrought electrodes and scaling up furnaces to 200 tons for round ingot production and to over 100 tons for slab production. They also are studying high-pressure melting of nitrogen-containing alloys to achieve savings in nickel and other costly alloying elements.

The widespread application of ESR enabled the Soviets to upgrade rapidly metal quality and manufacturing efficiency and to bypass a number of conventional steelmaking practices that had been developed highly by Western countries over many years. While it is easy to dismiss such widespread use of ESR as uneconomical by Western standards, the USSR found the ESR process to be both economically and technically the shortest and most efficient route from raw material to a finished industrial component or assembly.

An extensive, broad-based, and highly centralized research and development effort is maintained to support the Soviet ESR industry. This situation is considerably different from that in other countries, particularly the United States, where overall ESR research and development, while extensive, is decentralized and is conducted separately by individual steelmakers, independent and government laboratories, and equipment manufacturers. The center of Soviet research and development for "special electrometallurgy," which includes ESR, is the Institute of Electric Welding (IES) in Kiev. Other institutes, including the Baykov Institute of Metallurgy (IMET) in Moscow, are involved. The IMET and IES are responsible for obtaining foreign patent coverage and for marketing Soviet ESR technology in other countries.

The work at the IES includes fundamental refining and solidification studies, alloy development, equipment design and development, and programs directed at expanding the use of ESR. In an effort to rapidly identify optimum operating parameters for large furnaces, the IES, under V. F. Demochenko, is using mathematical modeling to determine temperature distributions in the ESR processing of large ESR ingots. The most significant development at the IES is the direct refinement of a "liquid metal" electrode to eliminate cast or wrought electrodes and permit continuous operation. (Present techniques for manufacturing the multiple long, slender, and straight electrodes needed for use in bifilar furnaces are costly. Electrode clamping, positioning, and drive control during operation and electrode change, if required, are complex and costly operations.)

To increase process efficiencies, the Soviets are studying the use of ESR techniques to produce multiple thin slab ingots using ingot withdrawal. This work is directed toward production of multiple small-section (up to 8 inch square) ingots by withdrawing ingots from a "tulip" mold.

For several years, the IES attempted to develop large ESR furnaces and, at one time, spoke of 330-ton ESR ingots for power-plant rotor shafts. Now, however, perhaps because of difficulties encountered when attempting to build units larger than 66 tons (see Table 1), they are concentrating on making large forgings by ESR welding a series of short 66-ton ESR ingots into a large composite ingot. Nevertheless, the Soviets have sold Poland's Nowatka Works a 220 ton furnace that will operate initially at a capacity of 165 tons.

TABLE 1 General Characteristics of Soviet ESR Furnaces.

Initial Date of Operation	Furnace Designation	Maximum Ingot Size		Electrical Circuit	Number of Electrodes
		Weight (tons)	Diameter (inches)		
1956	R909	0.7	12	1 ∅ (phase)	1
1958	DSS-1	1.1	--	3 ∅	3
1959	R951	1.7 [a]	--	1 ∅	1
	R951M	2.8 [a]	--	1 ∅	1
	OKB-905	2.8	--	1 ∅	1
1958-59	ESHP-2	2.8	24	3 ∅	3
1959-60	ESHP-10	11.0	48	3 ∅	3
1965-66	ESHP-10M	12.1-16.5	40	1 ∅	1
	OKB-906	3.9	--	3 ∅	3
1968	OKB-1065	3.9	22.4 x 22.4	1 ∅	1
1969	OKB-1111	44.0	60	3 ∅	3
1970	OKB-1111	66.0	--	--	--
1971	ESHR-200	220.0 [b]	--	3 ∅	3

Capacity figures for these units are unavailable.

[a] Round, square, and rectangular ingots.
[b] Reported experimental model at Paton Institute (probably never built).

2. Production Capacity

The USSR has a greater ESR productive capacity than the United States and produces a wider range of alloy products. While U.S. companies use ESR to produce superalloys, specialty steels, tool and die steels, classified nuclear alloys, and some high-strength steels, the Soviets produce these materials as well as a substantial quantity of the more common varieties of low-alloy steel for critical fabrications.

The Soviet electroslag casting product line includes valve bodies for nuclear reactors, large crank shafts, pressure vessels, and similar components; therefore, development and construction of the heating, forging, and machining equipment needed to produce these components conventionally has been reduced. Since the Soviets lack the vast forming and machining facilities as well as the required labor skills possessed by the United States, they explore using ESR to bypass conventional operations at every opportunity.

The Soviets introduced the ESR process on commercial scale at the Dneprospetal Steel Plant in 1958. By 1970, ESR furnaces were used in unspecified numbers in at least seven other steel plants located at Chelyabinsk, Elektrostal, Izhevsk, Volgograd, Zaporozh'ye, Zlatoust, and Novo Kramatorsk. The general characteristics of Soviet furnaces are presented in Table 1. Precise data on the number and individual capacity of the units are not available. The current total capacity (not including that of units in operation in Czechoslo Czechoslovakia, Poland, East Germany, and Yugoslavia) is estimated at under 200,000 tons per year.

Since 1965 several nations have purchased Soviet ESR furnaces and engineering, and although these installations have had their share of component performance problems (e.g., with ESR molds), good material generally has been produced in their equipment. Within the past 12 months, the Soviets placed a 44-ton ESR slab furnace in operation at Nippon Steel, Japan. This furnace, termed the ESR-40 (metric tons) by the Soviets, is of bifilar design and incorporates a moving collar mold. It also is designed to make round and square ingots (or to serve as a prototype for the larger furnaces). The USSR also has sold a hollow ingot ESR furnace to a Swedish steel company (SKF) but neither ingot size capability nor system design features have been disclosed.

E. The United States

1. Present Status

United States capacity for electroslag remelting is expected to reach 126,800 tons per year by the end of 1975 (Table 2). Any analysis of future market applications is very speculative; however, Table 3 presents ESR market projections extrapolated to the year 1980. The minimum expansion was derived by assuming an overall average growth rate of 10 percent per year, and the individual groups were adjusted on the basis of the current level of interest in

TABLE 2 Installed U.S. Electroslag Remelt Capacity Estimated for 1975.

Principal Type of Materials	Annual Remelt (tons)	Percent of Total
Tool and Die Steels	34,400	27.1
Carbon and Low-Alloy Steel	27,700	21.8
Superalloy	39,100	30.8
Stainless	25,600	20.3
TOTAL	126,800	100.0

TABLE 3 Projected Capacity Increase and Market Application for ESR Materials in the United States for the Year 1980.

Principal Type of Materials	Expansion of Installed Capacity at 10% (tons per year)	Potential Application (tons per year)
Tool and Die Steels	55,300	86,000
Carbon and Low-Alloy Steel	44,500	256,800
Superalloys and Specialty Alloys	62,890	91,000
Stainless Steels	41,450	66,200
TOTAL	204,140	500,000

each field and an assumed expansion of existing capacity. Potential market applications for ESR materials are considerably greater than projected capacity. This is especially the case in the iron-base materials category (including, for example, carbon and alloy steels, rotor steels, roll alloys, armor plate) and reflects the increasing requirement for more reliable, higher performance steels. ESR also is expected to be applied more widely to nickel- and cobalt-base alloys that presently are air-melted and cast into conventional ingots.

Economic considerations obviously will determine how much additional ESR capacity is realized. The market for carbon and low-alloy steels, while much greater than that for the other categories, is also the most sensitive to product cost and value analysis of materials with improved properties.

IV. GENERAL PROCESS DESCRIPTION

A. Principles of Electroslag Remelting Process

A schematic of the ESR process is shown in Figure 1 (page 8). The consumable electrode, usually produced by casting after primary melting of the alloy, is remelted into an ingot in a water-cooled crucible resting on a cooled base plate. Electrical connections are made to the base plate and the electrode and the power supply may be a variable tap transformer or, if direct current is used, a rectifier.

The process is in many ways similar to other consumable electrode remelting processes, particularly vacuum arc remelting, but it is unique in its way of generating heat. Heat is generated in the flux pool by resistance to the electric current (Joule heating) and it causes the electrode tip to melt and drip down through the superheated slag into a molten metal pool where it solidifies progressively and builds up an ingot. This heat also supplies the energy required to maintain the ingot and slag temperature gradients necessary for the desired solidification structure. As the metal melts on the electrode and droplets pass through the slag bath, an extremely large surface area is exposed to the hot, highly reactive slag.

Effective refining takes place during the formation of the liquid metal film on the electrode, the fall through the slag, and in the liquid metal pool. A continuous shell of solid slag which is present between water-cooled mold and solidifying ingot provides smooth ingot surfaces. A sound ingot with improved metallurgical characteristics is built up progressively by nearly vertical solidification. By suitable selection of fluxes, chemical reactions can be promoted or inhibited, and the removal of nonmetallic inclusions is enhanced by chemical extraction and physical flotation from the pool.

Good accounts of the development of the ESR process are available in the literature (Duckworth and Hoyle, 1969; Latash et al., 1971; Leybenzon and Tregubenko, 1963).

B. Basic Process Parameters

Numerous variations and modifications of the basic ESR process have been studied, proposed, and investigated, and some have reached industrial applications (see Table 6 for a list of the many process and equipment parameters affecting productivity [melt rate], power efficiency, refining capability, composition, and ingot structure control). The versatility of this refining process is evidenced by the several power modes and different electrical circuits that can be used. The choice between the principal power modes—alternating current (AC) and direct current (DC)—and their modifications is dictated by technological factors including electrical considerations and local conditions.

The two power modes do, however, have different metallurgical results, and small and medium size installations generally use single-phase AC power. Usually, one electrode is remelted into one crucible or two electrodes can be converted into one or two ingots (see section VII.A). While this power mode is simplest and lowest in cost, line imbalance and inherent inductive losses increase considerably with high power inputs, militating against its use when producing large ingots. To produce larger diameter ingots, electrically balanced systems employing three-phase AC and DC power can be used.

The DC and low-frequency power modes have no such electrical limitations and can be used for production of any ingot size as well as in polyelectrode remelting. As indicated in section VII.A (Table 26), the main drawbacks of these modes are higher cost of power supply and, possibly, lower operational reliability. Because electrochemical reactions occur in DC electroslag refining, chemical and metallurgical control are more complex than when AC is used. The severity of electrolytic reactions depends on the current density, which means that different refining results may be obtained in small and large ingots. DC is used primarily in converted VAR furnaces where reversed polarity (electrode positive) is preferred (Dewsnap and Schlatter, 1974).

Each system has features that must be balanced with its shortcomings. However, acceptable remelting results can be achieved with most power systems if proper slag compositions and appropriate melting parameters are employed. Electrical parameters are the most important and are determined primarily by the shape and depth of the metal pool, the slag system, and the power efficiency. The best compromise between metallurgical and economic requirements is often difficult to achieve since a shallow metal pool, which is desirable, is not compatible with the high melting rates needed for maximum productivity and ingot surface quality. Melting rates, which will be discussed in section V.C (Figure 6), vary with power input and power mode.

Heat is transferred into the ingot either directly from the slag or through the sensible heat in the metal drops falling from the electrode. Thus, the energy contained in the metal pool is dissipated by conduction into the surrounding water jacket and, as a result, a large temperature gradient is maintained through the solidifying zone. Local cooling rates in industrial ESR ingots typically are in the region of 2 to 9° F (1 to 5° C) sec^{-1} which leads to the fine dendritic structure commonly observed in these ingots. For many years, the high degree of chemical refining was attributed to the large slag/metal contact area resulting from the droplet mode of melting in the process. This assumption proved to be false, since the droplet residence time is exceedingly small. The main effect of the droplet mode is to bring the metal uniformly to the slag temperature through the high rate of heat transfer in the system. The sensible heat, thus acquired, constitutes the major part of the heat entering the ingot.

The thermal inefficiency of the freezing part of the ESR process is a necessary component of the ingot structure control. In order to maintain control of the solidification process, the directional heat flow in the solidifying region must be dictated largely by the externally imposed temperature gradient. The overall thermal efficiency of the process is, of course, approximately zero since the starting electrode and final ingot are at similar temperatures. However, the specific energy efficiency of the melting process can range from 1000 to 2000 kWhr per ton.

The structure of the solidifying ingot is determined by the melt rate, freezing rate, slag temperature, and direction of heat extraction. A low amperage, a high voltage, a deep slag pool, and a slow melting rate produce a thicker and better insulating skin of slag between ingot and mold that results in more vertical freezing. Conversely, radial solidification is promoted by high amperage, low voltage, a shallow slag pool, and a fast melt rate. Electrode/ingot geometry affects the heat distribution in the slag pool and, consequently, the size and shape of the metal pool.

The ingot is usually finer in structure and has less macro- and micro-segregation than an equivalent section of a conventionally cast ingot. These effects are to be expected from the solidification conditions and are common to all consumable electrode processes. The ESR ingot, however, also may contain all the solidification defects observed in other processes if the freezing has been performed under adverse conditions. Examples of these defects are freckles, which result from pool stirring, and microporosity, which results from gas evolution. Solidification of the ingot occurs against a smooth, thin skin of solidified slag. There is no spatter "crown" as in VAR and no chill freezing, as in electron beam (EB) or plasma arc melting (PAM). The ingot head contains a small cylindrical portion of liquid metal that progressively forms the new ingot surface. The extreme stability of this region provides the very high surface quality obtainable in ESR ingots.

C. Slags and Process Chemistry

The selection of a satisfactory slag or flux system is essential for the success of the remelting operation. The choice of slag is governed by physical (electrical conductivity, liquidus temperature, vapor pressure, viscosity, thermal capacity and conductivity, and interfacial tension) and chemical (thermodynamic properties, high temperature stability, and oxygen potential) considerations. The process generally is one of oxidation; hence, precautions are necessary to minimize the oxidation losses of alloying elements with high free energies of oxide formation (e.g., silicon, aluminum and titanium). The slag provides the source of heat because of its resistivity and controls the composition of the remelted alloy.

The slags are usually liquids in the system $CaF_2 + CaO + Al_2O_3$ and range widely in composition. Sometimes additions of SiO_2, TiO_2, and MgO are used. In general, CaF_2-based slags are chosen with additions of Al_2O_3 and CaO to increase the resistivity and promote sulfur removal, respectively. Alkali metal oxides and transition element oxides are to be avoided. Maximum desulfurization is achieved with slags with a basicity greater than 5. Care must be exercised in purchasing slags to ensure their freedom from unwanted trace elements (e.g., lead) that would transfer to the ingot during ESR. Impurity levels may require further restriction for certain alloy grades if high purity in the alloy is to be maintained. The quality of the fluorspar is particularly critical, and three grades are on the market:

1) Metallurgical quality with 80 percent CaF_2 minimum (generally used in steelmaking operations)
2) Ceramic quality with 95 to 96 percent CaF_2
3) Acid spar with 97 percent CaF_2 minimum and less than 1 percent SiO_2 (used for the production of hydrofluoric acid and derivatives)

Of these three qualities, only the acid grade is acceptable for ESR fluxes. Mexico supplies most of the fluorspar used in the United States but the availability of high-grade fluorspar is limited. Expected shortages of acid-grade fluorspar have stimulated research in Europe to develop fluxes with low fluorspar content for the ESR process (Pateisky et al., 1972). Additional development work should be initiated to drastically lower the CaF_2 level and to recycle ESR fluxes.

Slags without CaF_2 may be employed although their use on an industrial scale has been very limited. Slag compositions also are discussed in section V. The slag for a particular alloy must have the following properties:

1) Resistivity characteristics suitable for adequate heat generation
2) Liquidus temperature below that of the alloy
3) Composition such that the solid precipitating from the slag will form a suitable slag-skin for good ingot surface quality
4) Chemical compatibility with the alloy being melted

Most basic rules governing the selection of a slag composition to meet these criteria in a given case are known. The application of thermodynamic principles relating to stable equilibria and their relation to mass transfer kinetics has been demonstrated in a wide range of slag/metal combinations in ESR. The relation between slag-skin composition and ingot surface quality also has been shown but the details of the mechanisms that determine the reactions and heat flow remain substantially obscure despite significant research work. Hence, the choice of slag composition is based presently more on operating experience than on fundamental knowledge.

In terms of slag behavior, the choice of AC or DC power becomes significant. The electrochemical effects observed in DC operation (due to the current passing between electronic and ionic conductors at each slag/metal Faradaic interface) cause reactions between the slag and the metal in DC melting that are distinctly different from those in AC melting. In addition, the same reactions are associated with heating effects at these interfaces. These heating effects are responsible for the basic differences in melting efficiency between the AC, DC electrode positive, and DC electrode negative modes.

Chemical effects in ESR have been associated traditionally with refining (e.g., the relationship between the sulfur and oxygen content of the electrode and that observed in the ingot following a particular remelting practice is well defined). The electrochemical and thermodynamic reactions leading to refining also have been described reasonably well in terms of simplified theories. In free-machining steels, the knowledge of these reactions led to the development of techniques to retain sulfur in electroslag processing.

The refining step for oxygen generally is associated with the removal of oxide inclusions. The contact between slag and metal is extremely good in ESR and oxide inclusion compositions are generally highly soluble in the slags used. Oxides are removed by the solution step and, hence, the ingot oxygen content generally is lower than that of the electrode. In addition, since ingot segregation is lower than that found in conventional ingots, the size distribution of oxide inclusion tends toward sizes finer than those associated conventionally with a given oxygen level. The mechanism of inclusion removal has been the subject of some controversy. Essentially the postulate of inclusions floating from the ingot pool under buoyancy forces has been discredited, and most inclusions are thought to form from solutions in the ingot liquid metal pool by oxide precipitation at the liquid/solid interface.

Very extensive work has been performed in adjusting slag composition to control the remelted alloy composition. In the case of alloys containing titanium and aluminum, for example, the TiO_2 and Al_2O_3 contents of the slag are maintained at a ratio that is reasonably close to equilibrium with the desired alloy composition. The detailed chemical mechanisms underlying this technique are not well understood and significant controversy surrounds the efficacy of the method.

D. Basic Furnace Designs

The mechanical aspects of the basic ESR furnace design are rather simple. Many production furnaces are built on the column and crosshead principle, which provides ample working space and simplifies construction and operation. Other alternative designs are based on vacuum arc furnace concepts without the enclosed midsection and with a simplified control system. The three principal mold-ingot-electrode mechanical arrangements—static mold, moving mold, and ingot withdrawal system—are shown in Figure 2 (Schneider, 1969). The latter two normally employ electrode change operations to maximize ingot weight and productivity. The techniques differ in engineering detail but their effect on ingot structure is the same. The static mold system is the simplest and most reliable but least flexible design. To produce a reasonable ingot height with a single electrode, the electrode/ingot fill ratio must be about 0.6 or greater. With electrode change, smaller fill ratios and shorter electrodes that offer some advantages in electrode mold inventory but increase electrode manufacturing costs, can be used. Inductive losses and electrical control problems are

FIGURE 2 Schematics of Basic ESR Furnace Designs (from Schneider, 1969).

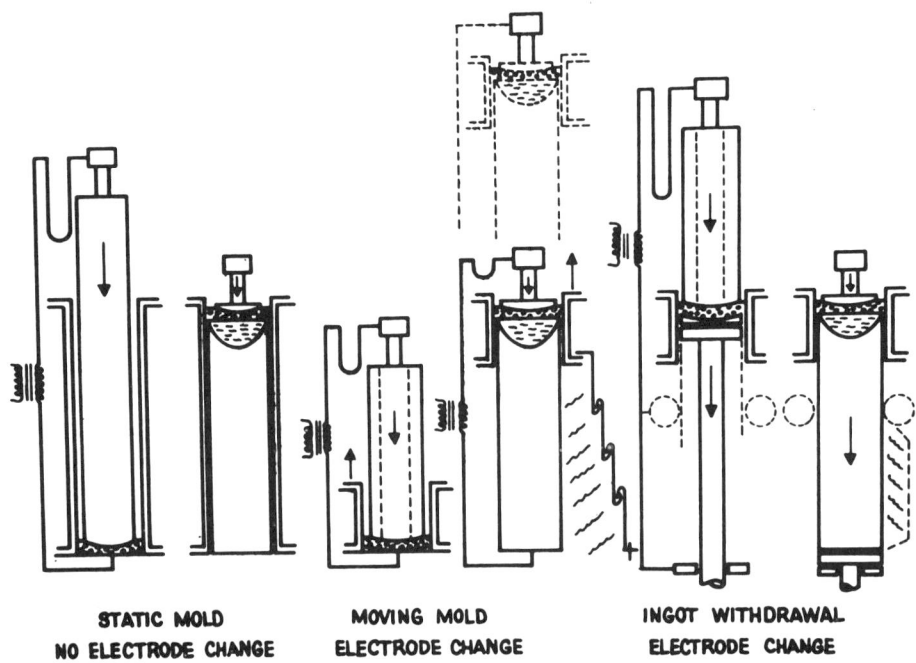

greater for single-electrode static-mold systems, but modern coaxial designs have eliminated these drawbacks. Moving mold and ingot withdrawal systems are mechanically more complex and costly, but they are self-stripping and extra stations therefore are not required. Heat shields, as indicated in Figure 2, can be used in both systems to keep thermal-sensitive ingots from cracking. The short mold systems provide ready access to the melt for visual control, temperature measurements, and chemistry sampling. Collar mold systems require restrictions on slag compositions imposed by the need to maintain a stable slag/ingot/crucible contact area. The major disadvantages are the poor surface quality of the ingots and micro- and macro-structural perturbations at the point where the electrode is changed, which may not be tolerable in some products.

The process normally is performed in open atmosphere which facilitates set-up and ingot extraction—a significant production advantage over other consumable electrode methods. The disadvantages lie in the process' sensitivity to atmospheric moisture and oxygen content and the evolution of fumes. The furnace atmosphere can be controlled using a dry-air or inert gas cover and the emissions can be controlled by using suitable extraction and filtration equipment.

ESR equipment technology has evolved differently in various parts of the world as process and equipment variables were evaluated but the three basic designs in operation today have universal acceptance. Depending upon capacity and productivity requirements as well as metallurgical aspects, various melt station arrangements are possible but the capital costs per annual ton are generally very similar for the different layouts. Other furnace designs, such as the bifilar ESR system, are described in section VI.B, and equipment technology is discussed in greater detail in section VII.A.

E. Energy Requirements

Consumable electrode remelting processes are electrically less efficient than primary melting processes because they remelt into water-cooled crucibles to obtain the metallurgically desirable characteristics of directional solidification. Power consumption tends to decrease as ingot section increases. A large slab-type ESR remelting facility requires approximately 1000 to 1500 kWhrs per ton of remelted product or approximately two to three times the 450 to 550 kWhrs required to produce a ton of liquid steel in a typical electric furnace shop. However, the total energy required to produce even the maximum projected demand for ESR materials represents a negligible increase in the overall annual energy requirements for U.S. steel production.

ESR requires more energy than VAR due to the additional heat transfer from the slag pool to the crucible and, as a melting process, has an overall electrical efficiency in the range of 20 to 45 percent (Machner, 1973). As shown in Table 4, the amount of power consumed by an ESR installation depends on the power mode used. It should be noted that the Soviet bifilar process has been claimed to have a significantly improved power efficiency but the data appear controversial.

TABLE 4 Effect of ESR Power Mode on Power Consumption.

	AC Single Phase	DC El- Polarity	DC El+ Polarity	Vacuum Arc Remelting
Melt Rate				
(pounds per minute)	8.0	9.8	6.5	7.5
(pounds per hour)	480	590	390	450
Spec. Power Consumption				
(kilowatt hours per pound)	0.49	0.39	0.58	0.36
(kilowatt hours per ton)	1079	860	1290	750
Oxygen Content				
(parts per million)	34	65	31	18
Sulfur Content				
(percent)	0.007	0.013	0.009	0.012

Electrode: 9 to 9.5 inch diameter, 25 parts per million oxygen, 0.013 percent sulfur. Crucible: 12 inch diameter. Steel Grade: M-2 high-speed steel. Power Input: 220 to 230 kilowatts. Flux Type: CaF_2-Al_2O_3-CaO. Slag Pool: 4 inches deep.

SOURCE: Dewsnap and Schlatter, 1974.

Several controllable variables affect the melting efficiency of the ESR process, and the electrode/ingot fill ratio is reported to have a considerable influence (Roberts, 1969). In the United States, fill ratios higher than those common in Europe and the Soviet Union are used. If limited power is available and an ingot or mold withdrawal system is employed, lower fill ratios are claimed to be advantageous. The slag cap thickness also has an effect on power consumption. The smallest slag cap consistent with good ingot quality is used to improve efficiency (Dewsnap and Schlatter, 1974) since deeper slag caps tend to produce a shallower metal pool at the same melt rate.

Additionally, the slag composition, specifically its electrical resistivity, is another important factor determining the efficiency of the ESR process. Further, ESR is a secondary (re-)melting process utilizing an electrode of an alloy that is made in a primary furnace and that already has absorbed a considerable amount of energy. In order to get the true energy picture, primary melting and electrode/ingot yield must be considered. Assuming that the average power consumption per metric ton of basic electrical steel is 680 kWhr and the average electrode yield is 90 percent, the "intrinsic" energy contained in the electrode is 760 kWhr per ton. The overall energy requirement for melting a ton of ESR steel therefore is approximately 1800 to 2000 kWhr, indicating that modern melting systems are much more energy-consuming than one would think (Barraclough, 1974).

F. **Economic Aspects**

The economics of the ESR process are of overriding importance and, in addition to local conditions and specific product requirements, usually dictate its applicability. It is, however, difficult to assess the true economics of the process because of the many auxiliary and intangible factors involved.

Cost comparisons often are made between VAR and ESR, but they depend strongly on the type of equipment and melting practice used, which explains the widely different figures quoted in the literature. The slightly lower capital cost of modern electroslag equipment in comparison to VAR installations is outweighed by higher operating costs due to lower power efficiency and considerable flux costs. Hence, the principal cost-sensitive areas in ESR melting are power consumption and fluxes. Both areas need improvement if the process is to be economically more attractive in the future.

The spent slag from the ESR process represents a relatively valuable by-product because of its CaF_2 content. Careful consideration must be given to recycling at least part of this slag, especially in those cases where maximum chemical refining during remelting is not required (i.e., where the principal emphasis is on solidification control). A major portion of spent slag has been recycled successfully in the electroslag remelting of certain tool steels. Another use of ESR slag is in the basic electric arc furnace as a CaF_2 substitute and general slag conditioner. With the growing shortage and increasing cost of high-quality fluorspar, recycling of the slag caps through crushing, blending, and, possibly, pyrometallurgical refining must be considered.

Economic studies of the ESR and VAR processes have revealed that the operating cost differential for modern furnaces specifically designed for ESR and VAR is very small (Wooding and Cerstvik, 1968), and the higher melt rates possible with ESR cannot shift the cost calculation sufficiently in favor of ESR. The aim is to recover the remelting cost in subsequent processing with significant yield improvements and elimination of operations. The marked upgrading of quality over the air melt product or more stringent customer requirements also may justify use of ESR.

The ESR yield of high-speed steel ingots is reported to be 10 to 15 percent higher than the air-melted yield of ingots of comparable size; however, the substantial cost of ESR ingot production is not offset completely by the increased yield of a costlier product (Kirk and Goodwin, 1973).

The strong effect of ingot size on the economics of ESR processing has been investigated for forged round billets of stainless steel and tool steels (Arwidson, 1973). The break-even ingot sizes for the same and different forging ratios are illustrated in Figure 3. The graphs indicate that relatively large ingot sizes in excess of 18 to 20 inches in diameter are required to compete economically with static cast ingots even when considering lower reduction ratios for ESR ingots. In comparison with usual static cast ingots for rolling, ESR does not appear to be profitable; however, ESR normally is considered cheaper than VAR because ESR ingots return higher yields in forging (Arwidson, 1973).

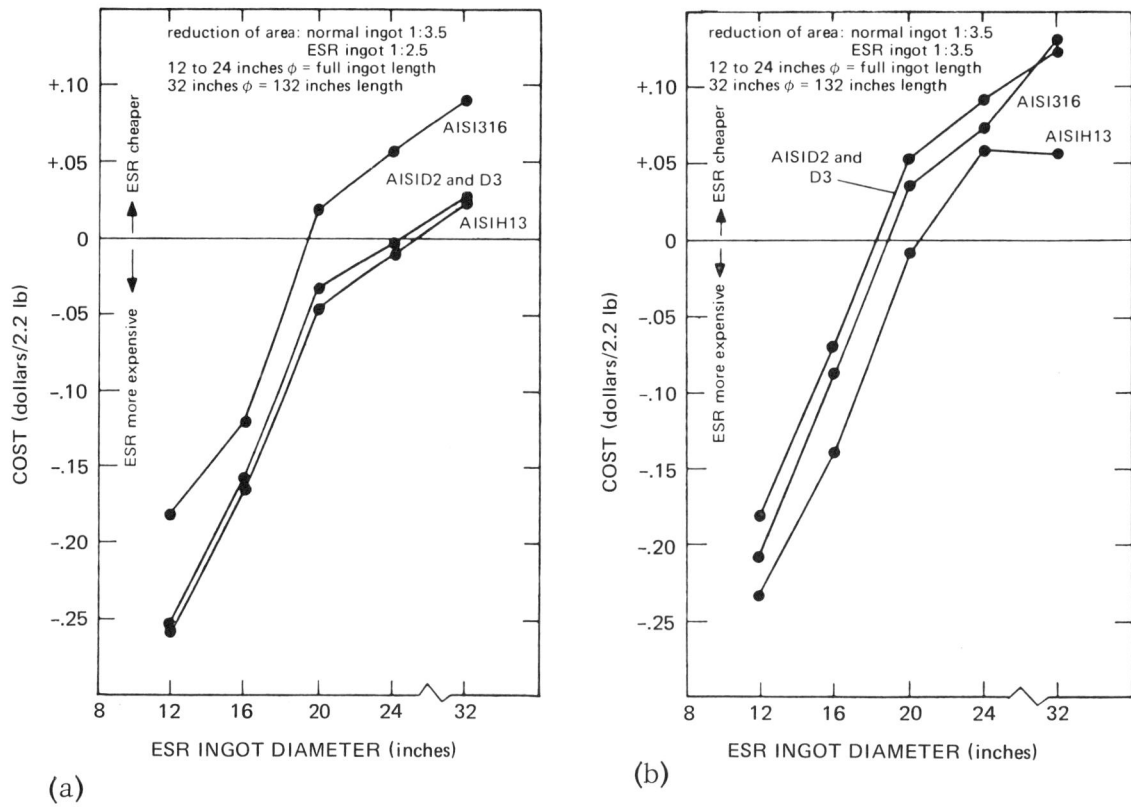

FIGURE 3 Comparison of Costs of Forged Round Billets from ESR and Normal Ingots Produced with (a) the Same Forging Reduction, and (b) a Lower Reduction Ratio for ESR Ingots (from Arwidson, 1973).

G. Shapes

An important feature of the ESR process is its capability to produce shaped ingots and castings. The shape of the ESR ingot is limited by the design of a cost-effective water-cooled copper mold and, generally, the greater the dimensions of a straight-sided mold, the more difficult is the job of producing a reasonably cost-efficient mold system. At present, the size limitations of the process are predominantly mold oriented rather than melting or power supply oriented. Nevertheless, the production of shapes by ESR is expected to effect significant cost savings (e.g., the reduction of scrap and elimination of processing steps could be achieved by employing an ESR hollow).

H. Environmental Considerations

The environmental effects of the ESR process are relatively easy to control. Most ESR facilities employ a closed water cooling system and dust and fumes are collected in a system employing either wet or dry techniques. One undesirable characteristic of the fumes from the process is the content of fluoride vapors and sulfur dioxide. The particulate matter may be collected readily in dry bag house systems.

Fluorides used in the ESR process constitute a health and environmental hazard. The degree of toxicity depends on the solubility of the fluoride in water wherein CaF_2 is the most insoluble (0.0016 g/100 cm³ H_2O at 19° C [67°F]) and BaF_2 has considerable solubility (0.12 g/100 cm³ H_2O at 25° C [77°F]). Harmful fumes, evolving from CaF_2 and oxide slags during fusion and remelting, contain SiF_4, TiF_4, SO_2, lead vapor, arsenic (As_2O_3), and CaF_2 (Duckworth and Hoyle, 1969). Calcium fluoride dust can cause fluorosis in humans and animals. It has been reported that fluorine volatization from high CaF_2 slags is by CaF_2 evaporation and HF formation, the latter increasing with higher moisture content (Schwerdtfeger and Klein, 1973). With SiO_2 present, SiF_4, rather than CaF_2 and HF, volatilization becomes predominant. To minimize fluorine emissions from ESR slags, dry materials and atmospheres must be used, and an efficient exhaust system must be installed over the slag prefusion furnace and around the opening of the remelting crucible. Generally, prefusion of the flux mixture generates considerably more fumes and dust than starting the process directly with solid slag. Fume emission during remelting is usually minimal and does not create any particular problem.

The major process waste is the used ESR flux, but a significant fraction can be recycled. The remaining waste fluxes, upon proper crushing, can be utilized readily as a replacement for fluorspar in almost any type of steelmaking. If the used fluxes must be disposed as solid waste, no harmful leaching characteristics appear to be associated with the mineral compounds because of their insolubility in water.

Environmental degradation as a result of electroslag melting of reactive, refractory, and heavy metals and alloys is minimal inasmuch as such operations usually are conducted in a closed furnace in an inert atmosphere. However, care must be taken in the melting operations to ensure that the reactive metals do not catch fire. The ESR process has the advantage of generating a relatively low level of noise.

V. ELECTROSLAG REMELTING TECHNOLOGIES

A. Electroslag Ingot Technology for Superalloys

The electroslag remelting of superalloys was the first significant use of the ESR process in the United States. The types of alloy that will be considered herein under the term "superalloy" include the heat-resisting, corrosion-resisting, and wear-resisting alloys, examples of which are shown in Table 5. These alloys are primarily nickel- and cobalt-base alloys although some iron-base alloys also may be included.

TABLE 5 Typical Superalloys Melted by ESR.

Alloy	Nickel-Base	Cobalt-Base	Iron-Base
Solid Solution	HASTELLOY B, C, C-276, G, S, X, F Inconel 600, 601, 625 Monel**, Monel 400**	HAYNES 25, 188, 31 HAYNES STELLITE 6B* MAR-M-302*, MAR-M-509*, X-45*	Inconel 800 16-25-6 RA-330 N-155
Precipitation Strengthened	Inconel 718, 713*, X-750, 901, IN-100** René 41, 77**, 95** Udimet 700 Astroloy** Rolls-Royce C-263 AF2-1DA** Waspaloy		A-286 Inconel 902, 706

* Small size ingots.
**Not remelted in production quantities.

1. Process Parameters

The various parameters associated with the electroslag remelting of superalloys are shown in Table 6. The most common furnace configuration, not including the USSR installations, is the single-phase AC mode. The DC mode seldom is used except in converted VAR units. In the processing of superalloys, significant disadvantages are encountered when remelting in the DC mode (Klein, 1970; Kajoika, 1973).

TABLE 6 Some Important Equipment and Process Variables in Electroslag Remelting of Superalloys

Factor	Variations
Power	AC
	DC Straight Polarity
	DC Reverse Polarity
	AC Poly Phase
	AC Bifilar
	AC Low Cycle
Equipment and Crucibles	Static Mold
	Movable Mold
	Withdrawing Ingot
	Spray Cooled
Electrodes	Variable Shape (round, square, etc.)
	Multiple Electrodes
	Electrode Change (normally in moving mold or withdrawing ingot)
Slags	Variable Slags Employed (depending upon alloy to be melted, desired melt rate, equipment employed, and technique chosen)
Starting	Dry Slag
	Arcing to Obtain Pool
	Exothermic
	Molten Slag
Special Techniques	Stirring of the Molten Pool
	Shaped Ingots
	Multiple Ingots
	Compositional Changes
	Composites

The particular slags that are employed for electroslag remelting of superalloys generally depend upon the alloy composition and the equipment that is used for the remelting operation. The slags are chosen primarily on the basis of electrical resistivity, vapor pressure, viscosity, slag liquidus, and chemical reactivity. Chemical reactivity considerations include the ability of the slag to remove inclusions, oxides, and sulfides as well as its inertness to the reactive elements (e.g., aluminum or titanium) in the metal. The other physical properties of the slag are related to melt rate and the surface of the ingot. (Slags and chemistry are discussed generally in section IV.)

For alloys containing reactive elements such as titanium, a small amount of TiO_2 commonly is added to the slag, especially in Europe. In simple thermodynamic terms, the following equation illustrates the expected effect of a titania addition to the slag:

$$\underline{Ti} + 2\underline{O} \rightleftarrows TiO_2$$

where the equilibrium constant

$$K = a_{TiO_2} / a_{Ti} a^2_O.$$

Thus, an addition of TiO_2 to the slag would raise the activity of titania in the slag and tend to drive the reaction to the left (i.e., less titanium would be lost to the slag before it reached equilibrium). A number of papers, written discussing these effects during electroslag remelting (Pateisky, 1972; Etienne and Mitchell, 1969), make it obvious that, at present, theory cannot be used to predict quantitatively the degree of reactive element loss. Significant basic work on interaction parameters and component activities in ionized slags is needed before such predictions can be made reliably. However, thermodynamics may be used qualitatively to point the way to a feasible solution and have been used extensively for superalloys. Unfortunately, most of these data remain proprietary. Since the ESR system is not a closed system, compositional control solely through slag metal equilibrium consideration is unsatisfactory.

One of the principal advantages of the ESR process, as will be discussed below, is its ability to remove certain inclusions (e.g., oxides and sulfides, but not nitrides) by dissolving them in the slag. This results in a material more workable (forgeable) and superior in properties to material processed by other methods.

a. Refining Reactions

To apply ESR techniques successfully to superalloys, reactions that may occur must be understood. Reactions that lead to refining, contamination, or loss of unwanted elements can be classified as mechanical (i.e., inclusion flotation), chemical, or electrochemical. Much work has been done to determine not only the mechanism but the site of inclusion removal during ESR (i.e., electrode tip, droplet, or molten slag-molten metal interface). The conclusion is that insufficient time exists for inclusions to be removed during the transfer of the droplet to the molten metal pool (Vachugov et al., 1967) and, therefore, inclusion removal most likely occurs at the electrode tip either by absorption of the inclusions into the slag through physical contact or by dissolution of the inclusion and subsequent chemical reaction of the molten metal in the molten slag. It has been shown for steels that approximately 50 percent of the inclusion reduction takes place at the electrode tip (Vachugor et al., 1967; Loyd et al., 1971).

Although the precise mechanism for the removal of inclusions during ESR is not known, a significant reduction in inclusion level occurs as evidenced by the data for HASTELLOY X (Figure 4). Here, the oxide inclusions after remelting were significantly less for ESR material than for comparable VAR material.

The removal of sulfide-type inclusions also is accomplished. Considerable desulfurization generally is experienced in alloys that are processed by electroslag remelting but not by vacuum arc remelting (Klein, 1972). The proposed reaction for the removal of sulfur is:

$$(M_x S_y) \text{ inc} \rightleftarrows x\underline{M} + y\underline{S}$$

$$\underline{S} + (O^=)_{slag} \quad (S^=)_{slag} + \underline{O}$$

Due to the very low sulfur levels normally present in superalloys, the sulfur activity in the slag always remains low and the mechanism of oxidation of the sulfur in the slag to $SO_2(g)$ is not an important consideration. This sulfur activity has some consequence in the discussion of reactive element control to be considered next.

b. <u>Reactive Element Control</u>

In electroslag remelting of gamma prime strengthened superalloys, the control of aluminum and titanium during remelting is critical. However, one important point should be remembered. Because the ESR process will remove inclusions effectively (i.e., Al_2O_3 and TiO_2, etc.), a small loss of chemically analyzed aluminum and/or titanium (which are tied up as oxides and/or nitrides) should be expected and is desirable during ESR.

The oxidative loss of reactive elements during electroslag remelting is determined by the oxygen potential in the slag that, in turn, is determined by the slag composition and the atmosphere above the slag. The presence of slag oxides that are reduced easily by the reactive elements in the metal results in losses of reactive elements, for example

$$2\,\underline{Al} + 3/2\,(SiO_2) \rightarrow (Al_2O_3) + 3/2\,\underline{Si}$$

Also, the presence of air ($P_{O_2} = 0.2$) above the slag bath will maintain a specific oxygen potential with respect to the slag. For molten slag starting, the slag commonly is deoxidized with aluminum before starting. If air is not excluded from the furnace annulus by using an inert atmosphere, the oxygen in the atmosphere quickly raises the oxygen potential to its value prior to aluminum deoxidation. The control of parts per million (ppm) concentrations of reactive elements requires careful determination of the correct electrode composition, slag system, and atmosphere as well as the proper choice of the remelting parameters.

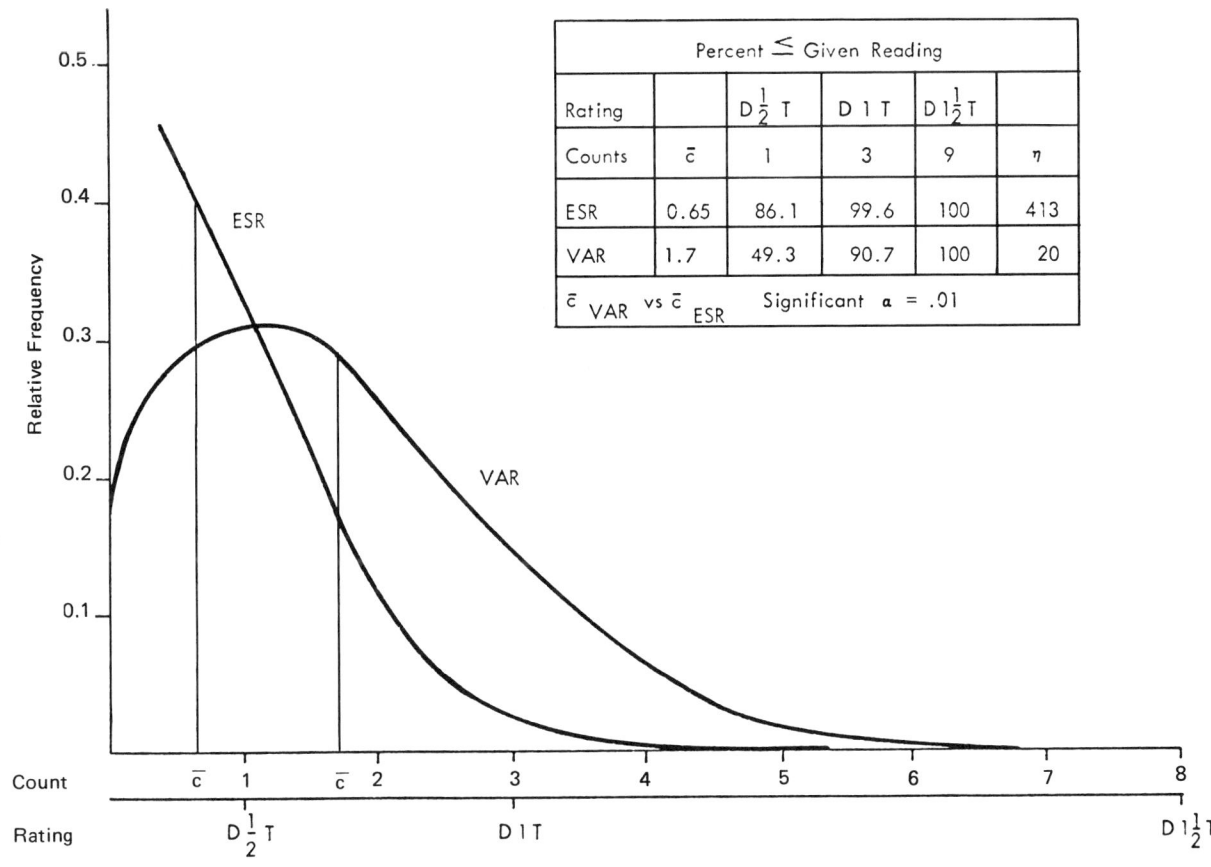

FIGURE 4 Distribution of Oxide Inclusion Ratings for 0.040- to 0.080-inch in Sheet of HASTELLOY X (from Kelley et al., 1971).

Electrochemical reactions always must be considered when using DC and, to a limited extent, when using AC due to the rectification that occurs (Klein, 1970; Hoyle and Duckworth, 1969; Medovar et al., 1963; Cameron et al., 1970). For superalloys, electrochemical reactions have been demonstrated to have a significant detrimental effect upon ingot quality (Klein, 1970). The results of a series of ESR runs using both the DC-straight polarity (DC-SP) and the AC mode on previously electroslag remelted materials are exhibited in Table 7 for HASTELLOY X and reveal that, when employing the DC mode, there is a loss of the easily oxidizable elements (aluminum and silicon) and an increase in oxygen. For the AC mode, there is no appreciable change. Similar results were obtained when using the DC-reverse polarity (DC-RP) mode. Thus, it was concluded from this experiment that the electroslag remelting of a superalloy should be done exclusively in AC furnaces if the highest quality ingot is to be achieved.

TABLE 7   Results of Chemical Analysis for Selected Elements Al, Si and O Compared to Electrode Values of 0.15, 0.45 and 0.002, Respectively.

Melt Mode	Position	Element	Slag Weight 4-3/4" Dia. Electrode			6" Diameter Electrode		
			10	14	18	10	14	18
AC	Butt end	Al	0.14	0.15	0.08	0.15	0.10	0.13
		Si	0.41	0.45	0.39	0.43	0.42	0.45
		O	0.002	0.002	0.002	<0.002	0.002	0.003
AC	Hot top	Al	0.10	0.08	0.08	0.10	0.11	0.11
		Si	0.44	0.42	0.21	0.42	0.43	0.42
		O	0.002	<0.002	0.002	0.003	0.002	0.003
DCSP[a]	Butt end	Al	<0.01	<0.01	0.03	0.07	0.11	0.11
		Si	0.23	0.025	0.21	0.39	0.27	0.39
		O	0.020	0.019	0.007	0.004	0.004	0.002
DCSP[a]	Hot top	Al	<0.01	<0.01	<0.01	0.05	0.06	0.08
		Si	0.20	0.017	0.23	0.36	0.35	0.38
		O	0.014	0.013	0.009	0.003	0.005	0.002
DCRP[b]	Butt end	Al				0.15		0.27
		Si				0.29		0.37
		O				<0.002		0.003
DCRP[b]	Hot top	Al				<0.01		0.12
		Si				0.036		0.33
		O				<0.002		0.007

[a] Direct current - straight polarity.
[b] Direct current - reverse polarity.

SOURCE: Klein, 1970.

2. Technical Advantages

For any processing sequence, the major goals are (a) increased alloy recovery and (b) improved alloy properties and reliability. Electroslag remelting of superalloys is successful because both of these aims are achievable.

a. Hot Workability

In the hot workability of a superalloy, the most critical step is the initial forging or breakdown of the ingot. Using Gleeble results that have been a good measure of forgeability, Figure 5 compares the workability of ESR and VAR ingots for Udimet 700. For the 4-inch-diameter ingots, the hot working range of ESR material effectively is double that of VAR material (Kelley 1969). Similar results have been noted for René 95 (Elliott et al., 1973).

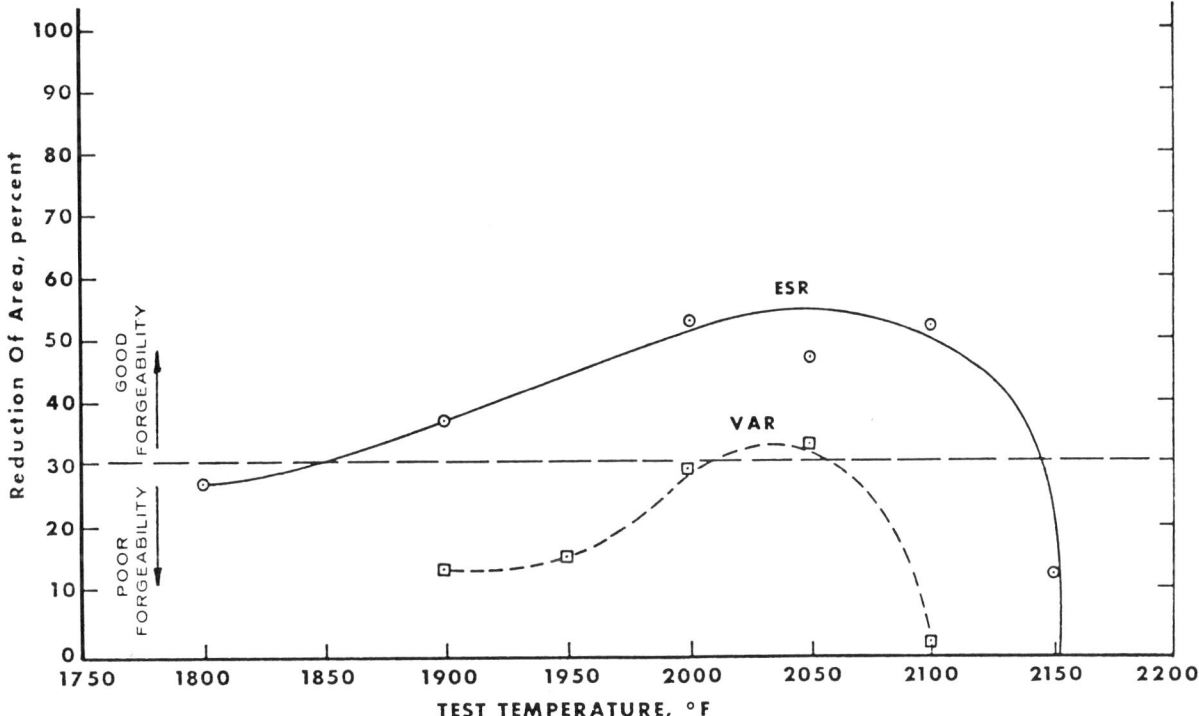

FIGURE 5 Gleeble Hot Ductility of Udimet 700 (from Kelley, 1969).

The improvement in the hot workability of ESR material over VAR material may be ascribed to a number of factors. ESR results in a significant decrease in inclusion content and sulfur level, thus eliminating severe hot workability problems. Another reason for increased metal recovery is the excellent surface quality of the ESR ingot. In addition, the ESR ingot characteristically has a relatively flat pool and, thus, a vertically oriented macrostructure that is more conducive to side forging than VAR ingot structure. Also, ESR significantly reduces ingot segregation. While not eliminated in all cases, freckling of superalloy ingots is minimized by the very stable ESR process. A better understanding of the freckling phenomena during remelting by ESR or VAR would benefit the superalloy industry significantly. The increased stability of the ESR process is a result of the heat generation by resistance heating of the slag as compared to heating by arc in VAR. VAR is susceptible to vaporization of volatile elements due to the use of low pressures that, in turn, can cause arc instability. In addition, in the electroslag remelting process, the molten slag thermally insulates the molten pool and dampens any abrupt change in power input. This is especially important when remelting alloys such as Inconel 718 that are quite freckle-prone.

b. Properties

The properties of ESR materials have been shown to be equivalent to, or better than, those of VAR material.

For HASTELLOY X, Table 8 indicates that the average properties of ESR material statistically are superior to VAR material and that the variation in the properties is much less (Kelley et al., 1971). Such an increase in material properties and reliability could lead to an increase in the design limits of ESR alloys because most structures are designed to a -2 or -3 $_\sigma$ limit. The improved chemical reproducibility of the ESR ingot compared to the VAR material also was shown in the same investigation (Kelley et al., 1971). For 7 out of 10 elements, the variation in the chemical analysis between the primary and remelt chemistry was less for the ESR material.

TABLE 9 Mechanical Properties of HASTELLOY X Bar.

Refining Method	Deviation	Tensile Strength (psi)	Yield Strength (psi)	Elongation (%)	Reduction in Area (%)
ESR	Average	+ 530	+ 270	+ 0.31	+ 1.10
	Standard	2,200	2,000	2.9	2.4
VAR	Average	- 580	- 270	- 0.25	- 1.14
	Standard	3,400	2,300	4.7	3.9
Average Properties (ESR and VAR)		107,980	48,400	50.0	54.3

SOURCE: Kelley et al., 1971.

For superalloys 600, 625, 750 and L605, the ESR process produces substantially the same properties as the VAR process. ESR porosity- and segregation-free ingots showed uniform composition in both transverse and longitudinal directions, with negligible compositional variation between the ladle analyses and the ingot. The worst field ratings for all types of non-metallic inclusions in ESR material were never greater than 1-1/2 thin (ASTM-E45-63, modified JK chart, method D) for any of the alloys. These results were comparable to VAR material. Routine tensile and stress rupture testing demonstrated comparable properties for ESR and VAR ingots of alloys 600, 625, 750 and L605 (Cook et al., 1971). Similar results also were reported for Astroloy (Klein et al., 1972).

René 41 and Udimet 700 have been electroslag remelted successfully (Bhat 1971). In both cases, ingot chemistry compared favorably with electrode chemistry, with 90 to 95 percent titanium and aluminum recoveries noted. Quaternary CaF_2-CaO-MgO-Al_2O_3 fluxes were used with both DC and AC power to melt ingots 4 to 8 inches in diameter and slabs 11 by 20 inches. For René 41, room-temperature tensile properties of plate forged from ESR ingots 7-1/4 inches in diameter were improved as a result of electroslag remelting; tensile strength at 1700° F (927° C) were somewhat lower than electrode material, although hot ductility was improved. Excellent hot workability was demonstrated. Recently, improved workability was noted with ESR AC, volatile-element-bearing, solution-strengthened, iron- and nickel-base superalloys using an ingot withdrawal technique (Elliott et al., 1973).

ESR processing of precipitation-strengthened alloys containing aluminum and titanium has been reported (Kelley et al., 1974). The results of the statistical analysis of the mechanical properties of Inconel 718 produced by ESR and VAR are exhibited in Table 9. The ESR material statistically had significantly better ductility than the VAR material. Similar results also were reported for René 41 and Waspaloy. The low-cycle fatigue life is expected to improve as a result of the increased ductility of the ESR material.

Three cast cobalt-base superalloys (MAR-M-302, MAR-M-509, and X-45) recently have been electroslag remelted into 3.5-inch-diameter ingots (Nafziger and Lincoln, 1974). The melts were made using AC single-phase power in 1/3 atmosphere backfilled helium (except for one X-45 melt) with different slag compositions. Vacuum arc melts were conducted on each alloy for comparison purposes. No significant improvement over the VAR ingots was noted in the globular oxide inclusions present in the ESR ingots with the exception of fewer thin inclusions in ESR MAR-M-509. Room- and elevated-temperature tensile and yield strengths of the ESR material generally were greater than those of the VAR material for the two MAR alloys. The tensile strength of ESR X-45 was less than that of VAR material below 1472° F (800° C). No trend in ductility, as a function of the melting technique, was noted. Stress rupture data showed that ESR MAR-M-302 withstood approximately 10 percent greater stress overall before rupture than the VAR material

TABLE 9 Comparison of AMS 5596C[a] Mechanical Property Test Results for ESR and VAR Flat Products of Solution Treated and Annealed HAYNES Alloy 718.

Mechanical Property	Overall Average	$\bar{X}_{ESR}$	$\bar{X}_{VAR}$	σ_{ESR}	σ_{VAR}	F	t	DF
R. T. UTS (ksi)	201.01	+0.38	-0.38	3.57	4.95	1.918[b]	1.209	173
R. T. 0.2% YS (ksi)	168.70	-0.32	+0.32	4.27	5.18	1.472[c]	-0.914	186
R. T. El (%)	17.71	+0.38	-0.38	0.89	0.89	1.013	5.812[b]	186
1200° F UTS (ksi)	163.86	+0.12	-0.12	5.41	5.06	1.142	0.318	186
1200° F 0.2% YS (ksi)	140.56	-0.40	+0.40	4.68	4.74	1.027	-1.168	186
1200° F El (%)	13.48	+0.65	-0.65	2.74	3.19	1.354[d]	2.987[c]	186
1200° F S/R Life (hrs)	4.67	+0.04	-0.04	0.39	0.40	1.036	1.476	186
1200° F S/R El (%)	1.93	+0.14	-0.14	0.40	0.38	1.104	5.094[b]	186

There were 93 ESR tests, with 51 heats and 93 ESR ingots; and 95 VAR tests, with 40 heats and 80 VAR ingots.

[a] 1775° F Solution Treatment.

[b] Rej. H_o: $\Delta \bar{X} = 0$ at 90% C.L.

[c] Rej. H_o: $\Delta \bar{X} = 0$ at 95% C.L.

[d] Rej. H_o: $\Delta \bar{X} = 0$ at 99% C.L.

SOURCE: Kelley et al., 1974.

at 1500, 1700 and 2000° F. For the MAR-M-509 ingots at 1500°F, the ESR material withstood only a slightly greater stress but for approximately a 50 percent longer time than the VAR material. However, at 2000°F, VAR material was considerably superior at lower stresses. The same trend was noted for the X-45 alloy at 1400 and 1900° F.

c. Novel Uses for ESR

The use of the ESR concept with minor process modifications offers a number of interesting possibilities. The introduction of such elements as nitrogen by electroslag remelting under a positive pressure of nitrogen could be quite beneficial. The concept has already been practiced for some austenitic stainless steels (Kubisch and Holzgruber, 1971), and some of the newer low-cost superalloys (e.g., Ferralium) rely on nitrogen additions. A series of superalloys is expected to be developed that requires the control of nitrogen at relatively high levels. To maintain these high levels during ESR, remelting under a positive pressure of nitrogen probably will be required.

Electrochemical reactions during the ESR process provide the opportunity to make alloy additions and to remove unwanted impurities. Most such electrochemical schemes that have been proposed involve using an auxiliary electrode or applying a DC bias to the electrode. Reportedly, oxygen and sulfur levels may be decreased by electrochemical reactions during ESR

(Holzgruber, 1969). In this method, the auxiliary electrode is connected with the main AC power supply but a rectifier is used to obtain the needed DC. For superalloys, the maintaining or introduction of reactive elements (e.g., rare earths) in small quantities by electrochemical means could be of significant importance both from the standpoint of cost savings and reliability.

3. Process Limitations

Most superalloys have been remelted successfully in production-size ingots by ESR. However, for certain alloys, remelt parameters must be chosen carefully to retain the reactive elements and eliminate segregation. A major limitation of the ESR process is that melt rate and heat input cannot be controlled independently. The heat input effectively controls the melt rate and one cannot, for example, go to a near-zero melt rate to allow for additional refining. For some superalloys, a relatively low melt rate may be desirable but this rate cannot be achieved without sacrificing surface quality in present ESR installations.

A second limitation of the process, as it is practiced today, is the inability to remove trace elements. Since ESR cannot be performed in a vacuum due to the volatile nature of the slag, the removal of trace elements, such as lead, is not possible. However, this limitation is not considered a major problem since most trace elements can be removed effectively during primary melting and, thus, this capability is not really needed.

An important deterrent to using material produced by the ESR process is that many aerospace industry specifications require VAR to be used. In general, these specifications were written before ESR was used extensively in this country, and changing specification requirements would require costly engine testing. In Europe, ESR material is accepted for this type of application.

One production method currently used for lower alloy materials employs an electrode change to increase ingot length without increasing the size of the electrode-holding portion of the ESR unit. Electrode change has been very controversial and is still not resolved. Theoretically, any such interruption must result in a change in the solidification process and, hence, segregation of the ingot. The extent of the segregation depends upon the alloy, section size, time required, and method of electrode change. The question to be resolved is how much segregation can be tolerated in an ESR ingot and that depends upon the intended use of the material. The use of superalloys in critical areas combined with the uncertainties associated with electrode change (i.e., can it always be accomplished in the required time) make the acceptability of electrode change doubtful with most superalloys.

4. Overseas Practices

The use of ESR material for critical aircraft applications has been accepted for a number of years in several countries. The three centers for ESR work in England are the Corporate Development Laboratories of the British Steel Corporation (BSC), Henry Wiggin, and Union Carbide Corporation (UCC) Glossop. At BSC, considerable work was done on slag compositions for superalloys although most of the work is unpublished and available only to members of the association. In France, reactive element control (e.g., titanium in iron-base superalloys) has been achieved in electroslag remelting (Antoine et al., 1967) and indicated that satisfactory properties, recovery, and hot ductility could be achieved using proper techniques.

ESR Monel metal produced in the Soviet Union showed an absence of hairline cracks and shrinkage porosity. Up to 10 percent manganese was lost with no other changes in chemistry (Lekarenko et al., 1964). Other nickel-base superalloys that were electroslag remelted in the Soviet Union showed improved high temperature ductility and impact strengths (as well as fewer inclusions) over air-melted material (Paton et al., 1967). Alloy compositions were not specified. Considerable work has been done in Germany on the effect of slag composition on reactive element control. Many European manufacturers tend to use low fill ratios while in the United States high fill ratios are employed universally.

5. Economic Aspects

For aerospace applications, the superalloys generally must be double melted; therefore, ESR that requires only one melt to produce an ingot of comparable quality competes directly with VAR and its double melts. Since the melting costs for the two processes are roughly equivalent, no economic penalties are incurred by choosing either ESR or VAR. In comparison to the VAR process, the ESR process results in a 5 to 10 percent yield increase as well as improved or equivalent properties.

When remelting is not specified for alloys, the economic aspect of ESR becomes more important. As stated above, the two incentives for using ESR are reducing production costs by increasing recovery and producing a superior product that may command a premium price. Thus, the decision concerning electroslag remelting of each specific class of alloys is considered on this basis.

6. Energy Requirements

No data on energy requirements for electroslag remelting of super-alloys are available but they should be comparable to the data reported for other alloys (Roberts, 1969; Holzgruber, 1968). One important consideration is the improved yield and properties of ESR material that may result in a decrease in the energy expended to produce a specific part capable of meeting particular requirements. For instance, better workability may increase the forging yield and decrease reheats, saving additional furnace time as well as labor. Also, the final part may be somewhat lighter if the effective properties of the ESR material are higher.

7. Shapes

The Air Force currently is funding an experimental program to develop aircraft-quality ESR superalloy hollows for use as starting stock in the production of turbine shafts, disks, and rings. For every pound of billet produced, an estimated 0.1 pound of material enters the aircraft turbine engine. Using ESR hollows as starting blanks for the production of seamless tubing for the chemical and nuclear industries also could offer a cost advantage.

8. Recommendations

1) A study of the physical chemistry of slag systems to identify the activities and activity coefficients of slag components during ESR should be undertaken to provide for a more quantitative thermodynamic treatment of the ESR process.

2) A comprehensive model of the ESR process should be developed. If an adequate model is developed, the scale-up from experimental to production size ingots would be facilitated. The model should include not only the normal heat-transfer fluid-flow considerations but also the solidification aspects of the process so that it would be possible to predict the occurrence of freckling in alloys to be remelted and, for example, determine the maximum size ingot that could be produced freckle-free along with the ESR parameters to be used. The possibility of electrode change during ESR also should be included.

3) A program to develop methods to control and remove trace elements during ESR should be undertaken. Since the ESR process results in a unique structure, the study should also focus on determining whether ESR materials will tolerate more or less trace elements than materials produced by other techniques.

B. Electroslag Ingot Technology for Carbon and Low-Alloy Steels

The electroslag remelting of carbon and low-alloy steels has encouraged the development of new products that offer material quality and mechanical properties significantly better than those of the products of conventional steel processing. However, insufficient economic leeway in basic costs prevents the application of ESR to most of these steel grades. Generally, the electroslag remelting of these steels involves much greater section sizes and weights than in higher alloyed systems.

The following steel grades are considered in this section:
1) AISI grades – carbon steels, 1006 to 1060; low-alloy steels, 4100 and 4300 series; plastic mold steels, P-20; and hot work die steels, H-11, H-12, and H-13.
2) ASTM-ASME grades (plate steels) – A203; A204; A302; A387 B, C, D, and E; A515; A516; A533; A537; A542; and A543.
3) AMS grade – 6359 (AISI 4340).
4) MIL-S grades – 16216 (HY-80/100); 18729C (AISI 4130); 24238; and 24371 (HY-130).
5) Gear steels.
6) Roll steels.
7) Rotor steels.

1. Process Parameters

In principle, the ESR process for carbon and low-alloy steels is similar to the ESR process for high-alloy grades. Generally, the incentives for remelting carbon and low-alloy steels are in achieving improved mechanical properties and uniformity in all directions of heavy sections. Particularly affected are impact strength or toughness, ductility or reduction in area, and fatigue properties. There is also widespread interest in improving the resistance to lamellar tearing in heavy-section steels that are stressed in the three principal dimensions. To obtain these characteristics, the electroslag process is used primarily to reduce oxide inclusions and sulfur content and also to obtain the beneficial structure that results from controlled directional solidification.

a. Scale of Process

One current application for these steels is the production of special quality steel plates (Swift, 1973). The present size consideration for the ESR units ranges from a remelted ingot cross section of 20 by 60 inches and weighing

10 tons to 30 by 80 inches and weighing approximately 30 tons. Such ESR facilities require about 3,000 kilowatts of electrical power and slag melting and handling facilities for 1 to 1.5 tons of molten slag.

For this scale of operation, appropriate high-quality flux raw materials of burnt lime, fluorspar, and alumina are bulk blended and remelted in a flux-melting furnace. An important consideration in preparing the molten flux is to maintain control of the dissolved carbon content since the linings of most large flux-melting furnaces are of carbon brick.

b. Molds

The full-size water-cooled copper molds for the remelting operation represent a significant cost item in the application of this process. Most experience to date has been with full-size or static molds that are of the straight or big-end-down tapered design, but strong current interest is being expressed in the short or "collar" type of mold.

c. Electrodes

The electrodes for remelting are typically cast in an electric furnace shop into ingot molds of the appropriate geometry to provide the desired electrode/ingot fill ratio. The electrode chemistry normally is controlled in the electric furnace shop by providing sufficient adjustments to the chemical composition to accommodate expected changes in the oxidizable elements during remelting. To minimize the hydrogen content of the electrode stock, vacuum degassing commonly is employed in their manufacture.

After casting, the electrodes may require the removal of excessive quantities of iron oxide (scale) on their surfaces. A holder or stub then is welded, with manual or semi-automatic welding equipment, to the electrode. This weld is necessary to attach the electrode to the supporting structure and to provide the path for the heavy electrical currents (up to 60,000 amperes) required in the remelting process.

d. ESR Operation

Typical practice in manufacturing these grades is to use a common starter plate material to facilitate the start-up operation and to ensure continuity of the electrical circuit. This starter plate has the same geometry as the mold and is placed on the copper base plate upon which the water-cooled copper mold is set. The thickness of the starter plate may range from 3/8 inch to 2 inches, depending upon grade and electrical conditions at start-up. Usually the flux is melted in a separate flux-melting facility and introduced into the mold cavity beneath the suspended electrode. After the molten flux is added and the current path established between the electrode and the molten flux, the remelting process begins.

The power is increased to some maximum level so that the remelting rate is equivalent to the solidification rate of the ingot, and the remelting process continues from 12 to 20 hours depending on size and weight of the electrode. Toward the end of the remelting process, it is usual practice to reduce the remelting rate for "hot topping" to provide sufficient molten metal to the solidifying and shrinking molten metal pool.

Upon completion of the remelting process, the water-cooled copper mold is stripped from the remelted ingot. Shrinkage of the remelted ingot during solidification provides sufficient clearance to permit removal of the mold from the solidified ingot. The ingot is then placed into slow-cooling containers prior to surface conditioning for rolling operations.

2. Technical Advantages

Carbon and low-alloy steel products that are produced by ESR offer several technical advantages over conventional air-melted materials. Generally, there is an improvement in product yield inasmuch as a greater percentage of the initial molten metal cast into electrodes may be processed into acceptable products. Also, higher yields may be obtained from the rolled product due to improved steel cleanliness and freedom from internal voids. Comparisons generally have not been made between VAR carbon and low-alloy steels and their ESR counterparts. Ultra-high-strength low-alloy steels (e.g., 4340, 300M, D6-AC) are produced by VAR and no attempts have been made to compare them with experimentally melted ESR materials.

a. Cleanliness

The internal cleanliness of ESR carbon and low-alloy steel products generally is much higher than that of similar products made in conventional electric furnaces. An indication of the degree of improvement of internal quality is that the ESR steels typically have cleanliness ratings evaluated by the J-K rating chart for ASTM E45 for vacuum melted steels. Typical microcleanliness results are reported in Table 10. Much more stringent specifications relative to ultrasonic testing and/or magnetic particle testing may be applied to heavy-section steels as a result of the improvement in cleanliness provided by this process.

TABLE 10 Inclusion Count Rating of Several Research Project Steels, Lectrefine® Processed.

Steel	Thickness (in.)	Ingot	Plate	Type A Sulfides Thin	Type A Sulfides Heavy	Type B Alumina Thin	Type B Alumina Heavy	Type C Silicates Thin	Type C Silicates Heavy	Type D Globular Oxides Thin	Type D Globular Oxides Heavy
ASTM A533-B	6	Top	1/4 thickness	1.0	0	0.3	0.3	0	0	1.3	1.5
		Bottom	1/4 thickness	0.5	0	1.0	0.3	0	0	1.5	1.5
HY-130	2	Top	1/4 thickness	0.5	0	1.5	1.0	0	0	1.5	1.0
		Bottom	1/4 thickness	0.3	0	1.0	0.3	0	0	1.5	1.0
ASTM A387-D	6-1/2	Top	1/4 thickness	1.0	0	0.5	0.3	0	0	1.5	1.0
		Bottom	1/4 thickness	0.3	0	0.5	0.5	0	0	1.5	0.3
HY-100	5	Top	1/4 thickness	1.0	0	1.0	1.0	0	0	1.5	1.0
		Bottom	1/4 thickness	0.5	0	1.0	0.5	0	0	1.5	1.0
	2	Top	1/4 thickness	1.0	0.5	1.5	1.5	0	0	1.0	1.0
		Bottom	1/4 thickness	1.0	0.5	1.5	1.0	0	0	1.5	1.0
HY-140	2	Top	1/4 thickness	0.5	0	1.0	0.5	0	0	1.5	1.0
		Bottom	1/4 thickness	0.7	0	0.7	0.3	0	0	1.5	0.7
AISI P-20	6-1/4	Top	Surface	0.3	0	0.5	0.3	0	0	1.0	1.0
			1/4 thickness	0.5	0	0.5	0.5	0	0	1.5	1.0
			1/2 thickness	1.0	0	1.0	0.3	0	0	1.5	1.0
		Bottom	Surface	0.3	0	0.3	0.3	0	0	1.5	1.0
			1/4 thickness	0.5	0	0.5	0.3	0	0	1.5	1.0
			1/2 thickness	0.3	0	0.5	0.3	0	0	1.5	1.0
AISI P-20	10-1/2	Top	1/4 thickness	1.0	0.5	0.5	0.3	0	0	1.5	1.0
		Bottom	1/4 thickness	0.5	0	1.0	0.3	0	0	1.5	1.0
4130	3	Top	1/4 thickness	1.5	0	1.5	0.5	0	0	1.5	1.0
		Bottom	1/4 thickness	0.3	0	0.5	0.3	0	0	1.3	1.0
4340	1	Top	1/4 thickness	1.3	0.5	1.5	0	0	0	1.5	1.0
		Bottom	1/4 thickness	0.3	0	0.5	0.5	0	0	1.5	1.0
	4	Top	1/4 thickness	1.5	1.0	1.5	0.5	0	0	1.5	1.0
		Bottom	1/4 thickness	1.0	0.5	1.0	0.3	0	0	1.5	1.0
4140	1	Top	1/4 thickness	2.0	1.0	1.0	1.0	0	0	1.5	1.0
		Bottom	1/4 thickness	1.0	0	0.5	0.3	0	0	1.0	0.5
	2	Top	1/4 thickness	1.2	1.0	1.0	0.7	0	0	1.5	1.0
		Bottom	1/4 thickness	0.4	0	1.2	0.4	0	0	1.5	0.7

All ratings were made by one operator using the method of ASTM E45, Plate III – the standard for "vacuum processed or other special quality steel." Worst field results are shown.

SOURCE: Lukens Steel Company, 1973.

b. Chemical Composition

When utilizing an appropriate slag, it is possible to produce steels with as low as 0.003 percent sulfur and 20 parts per million oxygen from electrodes containing 0.018 percent sulfur and 70 parts per million oxygen. Very low final sulfur and oxygen levels substantially improve the toughness characteristics of most steels. However, it is not always necessary or particularly beneficial to produce all steels in these grades to such low sulfur levels.

Chemical uniformity in a horizontal cross section obtained at any vertical position throughout the remelted ingot may be expected with ESR and is indicated in Table 11 (Lukens Steel Company, 1973). However, chemical variations in the vertical axis of the remelted ingot will not be subject to the same degree of homogenizing effect of the remelting operation. Variations in the composition of the electrode in the vertical direction can be transferred to the resultant ESR ingot.

The oxidizable elements must be protected from oxidation during the remelting process. Present practice employs an aluminum feeding system that precisely meters metallic aluminum to control the deoxidation of the ESR fluxes. Most commonly used ESR flux systems employ the ingredient Al_2O_3, which results in pickup of aluminum. The controlled addition of metallic aluminum does not upset the maintenance of the appropriate flux chemistry. In certain applications, harmful results have been reported from this aluminum pickup (Little et al., 1973; Kajoika et al., 1973). By utilizing special techniques for adding selected alloys during the remelting process, it is possible to change a remelted ingot composition from that of the electrode.

c. Mechanical Properties

Certain mechanical properties of steels are improved significantly by the ESR process while others remain unchanged. Yield strength, tensile strength, and hardness generally are unchanged but elongation, reduction in area, fracture behavior, and impact shelf energy are improved substantially. Also, properly remelted steels have mechanical properties with a high degree of uniformity in all directions of testing (i.e., the X, Y and Z axes). In addition, steels remelted with the appropriate flux chemistry are well suited for applications of very low service temperatures where impact strength and fracture toughness criteria are important.

Extensive mechanical property data from a comprehensive materials property evaluation for ASTM A533B plate steel are presented in Tables 12 through 20 (Swift and Gulya, 1973). These data provide an illustration of the improvement in ductility-dependent properties and also demonstrate a marked reduction in directional variation effects upon these properties.

TABLE 11 Chemical Analysis of Some Research Project Steels*.

Steel Type	Gage (in.)	Location Ingot	Plate	Element C	Mn	P	S	Si	Ni	Cr	Mo	V (Al)
4340	4	Top	Surface	0.41	0.78	0.008	0.004	0.31	1.67	0.83	0.27	NA
		Top	Quarterline	0.40	0.76	0.007	0.003	0.29	1.68	0.90	0.28	NA
		Top	Midgage	0.41	0.75	0.007	0.004	0.25	1.69	0.88	0.28	NA
		Top	Average	0.41	0.76	0.007	0.004	0.28	1.68	0.87	0.28	--
		Bottom	Surface	0.40	0.77	0.007	0.003	0.28	1.71	0.83	0.26	NA
		Bottom	Quarterline	0.40	0.76	0.004	0.003	0.30	1.63	0.81	0.26	NA
		Bottom	Midgage	0.40	0.72	0.006	0.002	0.29	1.63	0.80	0.26	NA
		Bottom	Average	0.40	0.75	0.006	0.003	0.29	1.66	0.81	0.26	--
		Overall	Average	0.40	0.76	0.005	0.003	0.29	1.67	0.84	0.27	--
		Total	Variation	0.01	0.06	0.004	0.002	0.06	0.08	0.10	0.02	--
4340	1	Top	Surface	0.40	0.75	0.008	0.003	0.30	1.66	0.85	0.27	NA
		Top	Midgage	0.41	0.75	0.008	0.004	0.23	1.70	0.86	0.27	NA
		Top	Average	0.41	0.75	0.008	0.004	0.27	1.68	0.86	0.27	--
		Bottom	Surface	0.42	0.81	0.010	0.004	0.29	1.67	0.91	0.28	NA
		Bottom	Midgage	0.41	0.76	0.010	0.003	0.29	1.67	0.90	0.28	NA
		Bottom	Average	0.42	0.79	0.010	0.004	0.29	1.67	0.91	0.28	--
		Overall	Average	0.41	0.77	0.009	0.004	0.27	1.67	0.87	0.28	--
		Total	Variation	0.02	0.06	0.002	0.001	0.07	0.04	0.06	0.01	--
P-20	10-1/2	Top	Surface	0.30	1.07	0.016	0.001	0.60	NA	1.81	0.41	0.078
		Top	Quarterline	0.32	1.06	0.015	0.001	0.59	NA	1.77	0.40	0.076
		Top	Midgage	0.31	1.03	0.016	0.002	0.56	NA	1.74	0.39	0.074
		Top	Average	0.31	1.04	0.016	0.001	0.58	--	1.77	0.40	0.076
		Bottom	Surface	0.30	1.04	0.015	0.003	0.58	NA	1.76	0.39	0.076
		Bottom	Quarterline	0.31	1.05	0.014	0.004	0.60	NA	1.78	0.39	0.077
		Bottom	Midgage	0.30	1.03	0.015	0.003	0.57	NA	1.75	0.38	0.074
		Bottom	Average	0.30	1.04	0.015	0.003	0.58	--	1.76	0.39	0.076
		Overall	Average	0.31	1.05	0.015	0.002	0.58	--	1.77	0.39	0.076
		Total	Variation	0.02	0.04	0.002	0.003	0.04	--	0.07	0.03	0.004

*IMPORTANT: The data on these pages have been obtained by testing at Lukens Steel Company. However, they are not to be considered a guarantee.

TABLE 11 Chemical Analysis of Some Research Project Steels (continued).

Steel Type	Gage (in.)	Location Ingot	Plate	Element C	Mn	P	S	Si	Ni	Cr	Mo	V (Al)
HY-140	2	Top	Surface	0.12	0.80	0.008	0.003	0.28	5.06	0.58	0.53	0.090
		Top	Quarterline	0.12	0.79	0.008	0.002	0.27	5.07	0.56	0.53	0.090
		Top	Midgage	0.13	0.79	0.008	0.004	0.26	5.10	0.56	0.54	0.090
			Average	0.13	0.79	0.008	0.003	0.27	5.08	0.57	0.53	0.090
		Bottom	Surface	0.13	0.79	0.008	0.001	0.28	5.05	0.55	0.53	0.090
		Bottom	Quarterline	0.13	0.79	0.008	0.003	0.28	5.10	0.56	0.53	0.090
		Bottom	Midgage	0.13	0.79	0.007	0.003	0.28	5.06	0.56	0.53	0.090
			Average	0.13	0.79	0.008	0.002	0.28	5.07	0.56	0.53	0.090
		Overall	Average	0.13	0.79	0.008	0.003	0.28	5.07	0.56	0.53	0.090
		Total	Variation	0.01	0.01	0.001	0.003	0.02	0.05	0.03	0.01	0.000
A516	5	Top	Surface	0.23	1.17	0.012	0.009	0.18	NA	NA	NA	(0.028)
		Top	Quarterline	0.22	1.17	0.010	0.009	0.19	NA	NA	NA	(0.032)
		Top	Midgage	0.22	1.20	0.012	0.009	0.20	NA	NA	NA	(0.032)
			Average	0.22	1.18	0.011	0.009	0.19	--	--	--	(0.031)
		Bottom	Surface	0.21	1.20	0.011	0.009	0.21	NA	NA	NA	(0.030)
		Bottom	Quarterline	0.23	1.22	0.010	0.010	0.21	NA	NA	NA	(0.033)
		Bottom	Midgage	0.22	1.22	0.012	0.011	0.21	NA	NA	NA	(0.039)
			Average	0.22	1.21	0.011	0.010	0.21	--	--	--	(0.034)
		Overall	Average	0.22	1.20	0.011	0.010	0.20	--	--	--	(0.035)
		Total	Variation	0.03	0.05	0.002	0.002	0.03	--	--	--	(0.011)
HY-100	2	Top	Surface	0.15	0.30	0.010	0.009	0.20	2.65	1.46	0.27	NA
		Top	Quarterline	0.15	0.30	0.011	0.006	0.20	2.65	1.45	0.28	NA
		Top	Midgage	0.15	0.29	0.010	0.006	0.19	2.65	1.45	0.28	NA
			Average	0.15	0.30	0.010	0.007	0.20	2.65	1.45	0.28	--
		Bottom	Surface	0.15	0.29	0.012	0.006	0.15	2.61	1.45	0.27	NA
		Bottom	Quarterline	0.14	0.28	0.010	0.005	0.15	2.58	1.43	0.27	NA
		Bottom	Midgage	0.15	0.29	0.011	0.006	0.15	2.60	1.45	0.27	--
			Average	0.15	0.29	0.011	0.006	0.15	2.60	1.44	0.27	--
		Overall	Average	0.15	0.29	0.011	0.006	0.17	2.62	1.45	0.27	--
		Total	Variation	0.01	0.02	0.002	0.004	0.05	0.07	0.03	0.01	--

SOURCE: Lukens Steel Company, 1973.

TABLE 12 Chemical Analysis in Weight Percent of Lectrefine® A533B (Heat R0048).

Element	Weight Percent
C	0.24
Mn	1.42
P	0.011
S	0.004
Si	0.14
Cu	0.10
Ni	0.52
Cr	0.11
Mo	0.59
Al	0.024

SOURCE: Swift and Gulya, 1973 (reprinted by permission).

TABLE 13 Inclusion Count – JK Ratings per ASTM E-45 Plate III (Heat R0048)

	A_T	A_H	B_T	B_H	C_T	C_H	D_T	D_H
Top Midwidth of Plate	1	0	0.3	0.3	0	0	1.3	1.5
Bottom Midwidth of Plate	0.5	0	1	0.3	0	0	1.5	1.5

SOURCE: Swift and Gulya, 1973.

TABLE 14 Room Temperature Tensile Properties of the Quarterline of 6-inch Gauge Lectrefine® A533B (Heat R0048).

Orientation	0.2% Yield Strength (ksi)	Ultimate Tensile Strength (ksi)	% Reduction in Area	% Elongation (2-inch gauge length)
Longitudinal	65.5	88.8	70.9	25.5
Transverse	67.8	90.8	71.3	25.8
Thru Gauge (Midgauge)	65.0	89.2	61.6	25.0

Heat Treatment: 1650° F – 3/4 hours per inch – WQ; 1280° F – 3/4 hours per inch – AC; 1150° F – 40 hours – AC.

SOURCE: Swift and Gulya, 1973 (reprinted by permission).

TABLE 15 Tensile Data for Electroslag Remelted A533B (Heat R0048)

Temp. (°F)	0.2% YS (ksi)	% RT YS	UTS (ksi)	% RT UTS	Y/T	% RA	% E (1" gl)
RT	66.2	--	90.2	--	0.73	73.2	25.5
0	70.1	106	96.9	107	0.72	72.1	26.5
-50	72.1	109	99.6	110	0.73	72.3	29.0
-125	82.6	125	106.9	118	0.77	69.2	27.6
-225[a]	105.9	160	119.3	132	0.89	64.8	27.7
-315	143.0	216	149.3	166	0.96	[b]	[b]

[a] One test, all others are average of two tests.
[b] Broke on punch marks.

SOURCE: Swift and Gulya, 1973 (reprinted by permission).

TABLE 16 Summary of Charpy V-Notch Impact Properties of the Quarterline of 6-inch Gauge Lectrefine® A533B (Heat R0048).

Orientation	FATT[a] (°F)	TT$_{40}$[b] (°F)	E$_{50}$[c] (ft-lbs)	E$_{-10}$[d] (ft-lbs)
Longitudinal	50	-30	100	55
Transverse	40	-30	100	60
Thru Gauge (Midgauge)	45	-5	70	40

Heat Treatment: 1650°F - 3/4 hours per inch - WQ; 1280°F - 3/4 hours per inch - AC; 1150°F - 40 hours - AC.

[a] FATT = Fracture Appearance Transition Temperature.

[b] TT$_{40}$ = Transition Temperature at 40 foot pounds.

[c] E$_{50}$ = Energy at 50°F test temperature.

[d] E$_{-10}$ = Energy at -10°F test temperature.

SOURCE: Swift and Gulya, 1973 (reprinted by permission).

TABLE 17 Fatigue Data for Transverse Quarterline of 6-inch Gauge Lectrefine® A533B (Heat R0048, Low Cyclic Rate).

Total Strain Range (ϵ_T) (%)	Plastic Strain Range (ϵ_P) (%)	Cyclic Rate (CPM)	Cycles to 1/16" Crack	Cycles to Failure
2.39	1.54	103	350	27,900 [a]
2.37	1.52	91	820	1,646
2.09	1.24	60	1,100	1,433
1.83	0.978	57	1,200	1,880
1.51	0.658	42	2,200	3,151
1.40	0.548	57	1,900	2,883
1.15	0.298	64	3,100	5,993
1.10	0.248	77	3,100	102,120 [a]
0.536	[b]	49	20,700	41,395
0.520	[b]	57	23,500	80,200
0.356	[a]	107	133,300	198,425
0.326	[a]	102	578,300	847,170
0.300	[a]	102	590,000	669,050
0.284	[a]	102	3,016,300	3,317,700
0.250	[a]	99	--	10,000,000

Total elastic strain = $\dfrac{\sigma_{YS}}{30 \times 10^6}$ + 2000 $\mu\epsilon$ = 4260 $\mu\epsilon$ = total elastic strain at yielding.

Plastic strain range = $\epsilon_T - 2\epsilon_Y$.

Heat treatment: 1650° F - 3/4 hours per inch - WQ; 1280° F - 3/4 hours per inch - AC; 1150° F - 40 hours - AC.

[a] See text for explanation of anomalies.

[b] Strain range within elastic limits of material.

SOURCE: Lukens Steel Company, unpublished data.

TABLE 18 Fatigue Data for Transverse Quarterline of 6-inch Gauge Lectrefine® A533B (Heat R0048, High Cyclic Rate).

Total Strain Range (ϵ_T) (%)	Plastic Strain Range (ϵ_P) (%)	Cyclic Rate (CPM)	Cycles to 1/16" Crack	Cycles to Failure
2.36	1.51	1,401	a	1,401
2.10	1.25	1,091	a	1,418
2.00	1.15	1,412	a	1,483
1.44	0.588	1,566	1,900	2,663
1.39	0.538	1,643	2,000	2,300
1.29	0.438	1,501	a	3,163
1.17	0.318	1,634	2,700	3,922
0.524	b	1,982	20,000	30,326
0.497	b	1,918	26,500	37,590
0.352	b	1,565	253,500	324,500
0.332	b	1,506	a	614,600
0.302	b	1,519	1,033,000	1,246,640
0.285	b	1,519	a	1,691,200
0.250	b	1,522	c	15,118,900

Total elastic strain = $\dfrac{\sigma_{YS}}{30 \times 10^6}$ + 2000 $\mu\epsilon$ = 4260 $\mu\epsilon$ = total elastic strain at yielding.

Plastic strain range = $\epsilon_T - 2\epsilon_Y$.

Heat treatment: 1650° F – 3/4 hours per inch – WQ; 1280° F – 3/4 hours per inch – AC; 1150° F – 40 hours – AC.

[a] Not determined.

[b] Strain range within elastic limits of material.

[c] No cracks observed after cycles noted.

SOURCE: Lukens Steel Company, unpublished data.

TABLE 19 Fatigue Crack Growth Rate ($\frac{da}{dN}$) for Electroslag Remelted Lectrefine® A533B (Heat R0048)

Fatigue Stress Intensity (ΔK_f) (ksi √in.)	Crack Growth Rate ($\frac{da}{dN}$) (microinches/cycle)
26.4	0.31
27.3	0.67
27.7	0.38
28.3	0.45
28.7	0.57
33.6	2.50
34.1	1.74

SOURCE: Lukens Steel Company, unpublished data.

TABLE 20 Fracture Toughness Data for Electroslag Remelted Lectrefine® A533B (Heat R0048).

Temp. (°F)	Yield Strength (ksi)	K_f (ksi √in.)	K_c (ksi √in.)	$0.6\left(\frac{\sigma YS_1}{\sigma YS_2}\right)K_c$ (ksi √in.)	$2.5\left(\frac{K_c}{\sigma YS}\right)^2$ (in.)	B (in.)	Reason for Invalidity
-300	140	28.1	36.0	10.6	0.165	1.983	a
-200	100	30.9	45.5	18.1	0.518	1.999	a
-100	79.0	36.0	85.2	42.8	2.910	2.000	b
		30.3	50.8	25.4	1.04	1.984	a
-75	74.9	36.5	91.8	48.7	3.75	1.974	b
		31.3	75.1	39.8	2.51	1.998	b
		28.9	83.5	44.3	3.10	1.994	b
-50	72.1	29.8	71.4	39.3	2.45	1.980	b

[a] $K_f > 0.6\left(\frac{\sigma YS_1}{\sigma YS_2}\right)K_Q$; $K_Q = K_c$.

[b] $B < 2.5\left(\frac{K_Q}{\sigma YS}\right)^2$; $K_Q = K_c$.

SOURCE: Swift and Gulya, 1973.

3. Process Limitations

a. Melt Rate

The melting rate of the ESR process is very low when compared with contenvional steelmaking processes because it is limited by the solidification rate of the molten pool. If the remelting operation is conducted at a substantially faster rate than the solidification rate of the molten pool, a larger molten pool results with the solidification pattern approximating that of conventional ingot casting. Therefore, one of the major cost considerations associated with ESR is its relatively low productivity per unit.

b. Power Supplies

Careful consideration must be given to the characteristics of the primary and secondary electrical circuits in large ESR power supplies since they generally employ a single-phase connection to the power line for economy reasons. This tends to distort the phase balance and can produce undesirable perturbations in a company's plantwide electrical distribution system if the single phase load represents a sizeable portion of the total plant load.

c. Molds

Although the available U.S. technology in ESR slab-type molds is believed to be advanced (i.e., relative to most other ESR producing countries), the cost associated with the operation of large ESR static molds appears very high, and the large, slab-type molds are a most serious limitation to size scale-up of the process. More applied technology in this field is necessary to advance ESR mold technology and to lower costs substantially.

d. Hydrogen

Hydrogen and its effects are one of the most serious process limitations relative to the widespread application of electroslag remelting of carbon and low-alloy steels. The pickup of hydrogen during ESR is recognized as one of the predominant metallurgical problems associated with this process, especially in locations of high humidity. The absolute hydrogen content that may be tolerated in various grades is not well defined at present. The effect of hydrogen is magnified by increasing section size and probably is related to the absolute inclusion content. Much of the work on hydrogen control is performed under proprietary investigations.

4. Economic Aspects

The economic aspects of the ESR process for carbon and low-alloy steels are much more critical than those of more expensive materials. As previously mentioned, the competition is very keen; therefore, in view of the high capital charges per ton of product, relatively high utility costs, and need for costly flux materials, a decision on incorporating the ESR process requires careful study of its operation and costs. Electrical energy consumption in conjunction with the low operating rate in tons per hour makes ESR operating cost per ton substantially higher than that associated with electric furnace melting. Therefore, a slight improvement in yield via ESR is insufficient economic incentive to produce these steels by other than conventional steel processing. The principal areas for cost improvement are:
1) Reduction in flux cost
2) Reduction in flux melting cost
3) Reduction in ESR mold cost
4) More effective utilization of electrical energy
5) Lower electrode manufacturing cost

A meaningful economic analysis must consider the additional value obtainable from steel products with improved mechanical and chemical properties and should be based upon the cost effectiveness of a given material requirement for individual applications. The use of ESR materials becomes more attractive economically the more stringent the quality and/or the higher the mechanical properties requirements.

5. Shapes

Round, rectangular, and square cross sections are typical carbon and low-alloy steel shapes being produced in the ESR process. The production of hollow rounds has generated some interest but not enough for viable commercial development at the present time. The cost of conventional ESR molds of rectangular cross sections approximately 30 by 80 inches may exceed $100,000; therefore, sufficient tonnage of a given specialty shape is required to justify the development and expense of molds with an unusual geometry.

6. Conclusions and Recommendations

a. Conclusions

(1) Hydrogen pickup in electroslag remelting of carbon and low-alloy steels is a major metallurgical problem and its impact has been recognized only recently. Currently, several manufacturers are conducting proprietary investigations to alleviate this problem.

(2) The ESR process basically is solidification-rate limited; therefore, present remelting rates cannot be increased substantially. Its low productivity, relative to air-melt steelmaking practices, leads to high capital costs per ton. Because electrical energy represents a substantial portion of the remelting cost, a reduction in electrical usage would represent a potential decrease in operating costs.

(3) The basic problem with large straight-sided molds is the distortion of the copper when exposed to the thermal gradient over long time periods. The maximum slab cross section size currently in production is approximately 30 by 60 inches, with a section size of 30 by 80 inches projected in the near future. With the development of larger straight-sided molds, operating difficulties have been increasing. The initial cost and maintenance expenses of these molds are a substantial portion of the total operating costs for these steels in large ingot sizes.

(4) Operating practices for large ESR furnaces, over about 20 tons capacity, generally require a molten slag to initiate the remelting process. The operation and maintenance of a slag-melting furnace also represent substantial cost items for this process.

(5) Solidification shrinkage tends to be an increasing problem with increasing section size. The greater the ingot dimension in the horizontal plane, the greater is the solidification shrinkage during hot topping. Because such shrinkage may lead to metal breakouts during termination of the remelting process, it is a serious problem in hot topping larger ingot sizes.

b. Recommendations for Research and Development

(1) Research and development for carbon and low-alloy steels should be initiated to control hydrogen at acceptable levels.

(2) Studies should be conducted to substantially reduce the kilowatt hours required per ton for the ESR process. One possible alternative may be the development of a "molten electrode."

(3) ESR molds should be developed that have reasonable manufacturing costs and a configuration that remains stable for long periods.

(4) Studies should be initiated to develop flux-melting facilities at substantially reduced costs and to permit the recycling of maximum amounts of these fluxes.

(5) Activity should be continued on the utilization of a DC bias potential on AC systems to promote electrochemical reactions during remelting.

(6) Shrinkage characteristics of large sections should be analyzed relative to residual stresses in resultant ingots so that effective techniques for hot topping large ingots may be developed.

C. Electroslag Ingot Technology for Stainless and High-Alloy Steels

The ESR process has found significant application in the manufacture of certain types of stainless and high-alloy tool steels that are difficult to melt and must meet specific quality requirements. These alloy steels cover a broad spectrum of iron-base compositions ranging from a total alloy content of approximately 10 percent up to 45 percent in complex special stainless alloys. Principal steel groups included in this section of ESR ingot technology are:
1) Stainless steels − ferritic stainless grades, 430* and 26Cr-2Mo*; martensitic stainless grades, 410, 431, 440A, 440B, 440C, and super 12 percent chromium steels (Jethete); austenitic stainless grades, 304, 316, 329, 321, 347 with titanium and niobium additions; and higher manganese austenitic grades, 202 with 8 percent manganese*, 216, nitronic stainless steels, valve steels, nitrogen added up to 0.45 percent; and precipitation hardenable stainless grades (not including high nickel austenitic heat-resisting alloys − see section V.A), 17-4PH, AM-355*, PH-13-8 Mo*, and semi-austenitic types.
2) High-alloy high-strength steels − maraging steel (18 Ni) grades, 200, 250, and 300; HP 9-4 Ni-Co steels, 0.20 percent C and 0.30 percent C; HY-180, 10 percent Ni steel; matrix steels (high-speed steel base)*; and multiphase Ni-Co alloys, MP35N*.

---

* These alloy steel grades have been produced by ESR only in experimental or pilot lots.

3) High-alloy tool steels – high-speed tool steels, Mo and W types; hot work tool steels (not including medium alloy 5 percent Cr type hot work tool steels – see section V.B), H20 series (W types), and H40 series (Mo types); and cold work tool steels, high Cr, D series, A-type steels.

For high-alloy steels, the ESR process always is considered a secondary remelting/refining process that is aimed at specific technical and economic benefits; it has never succeeded as a primary or "alloy making" process. Frequently, electroslag remelting of the above steels is used primarily as a "solidification tool" to produce ingots with controlled structures and superior characteristics that enhance further processing and final properties. In these cases, the chemical and physical refining during remelting are of secondary importance. In other applications, however, ESR is used to attain very specific quality improvements such as desulfurization and inclusion removal. Features like process flexibility and versatility also may be attractive for particular applications of the ESR process. A number of specialty alloys (e.g., electrical, magnetic, and expansion) have been melted successfully by ESR and the process is being used increasingly to meet the particular requirements of these alloys.

1. Process Parameters

The viability of the ESR process for high-alloy steels depends upon whether the melting costs can be offset by the metallurgical and yield benefits that may be realized. These benefits depend strongly on the reliable control of the metallurgical, physical, and chemical variables underlaying the ESR process.

In principle, the metallurgical quality of the product is controlled by the following factors:
1) Slag temperature
2) Melt rate
3) Solidification conditions

Electrical and physical conditions may be varied within limits to achieve this control. Slag temperature and melt rate are governed largely by the power input, but solidification conditions are considerably more involved. Figure 6 presents the relationship between ingot size, power requirements, and melt rates for high-alloy tool steels (Schlatter, 1974). The AC single-phase power mode is preferred for remelting high-alloy specialty steels because a wider latitude is provided in the selection of remelting parameters. Remelting by the AC bifilar setup appears acceptable only for segregation-insensitive alloy steels (e.g., austenitic stainless steels). A more realistic appraisal of slag refining reactions could be developed with a continuous temperature measuring device in the slag cap which would permit metallurgical control by temperature

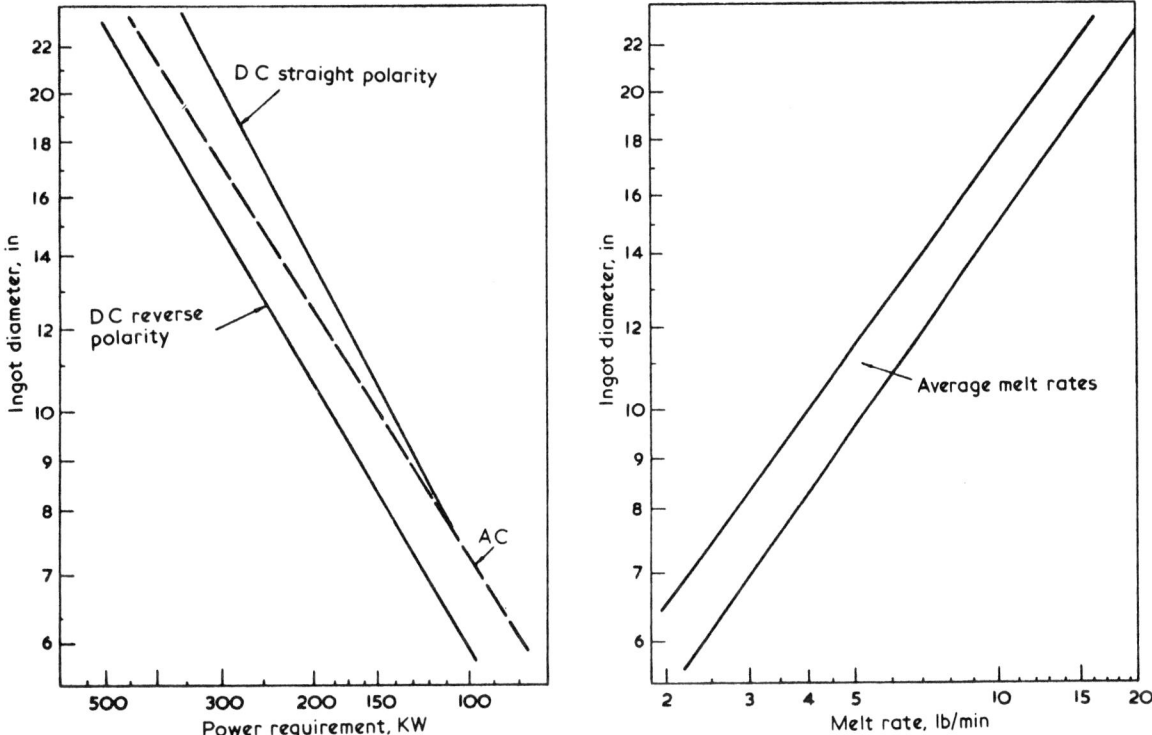

FIGURE 6 Power Requirements and Melt Rates versus Ingot Size for Electroslag Remelting of Alloy Tool Steels – Effect of Power Mode on Average Power Requirements (from Schlatter, 1974).

as an alternative to the present empirical approach based on power, current, and voltage. Conditions for heat extraction are very complex because of the slag skin around the ingot and the formation of a shrinkage gap between slag and crucible wall. Fortunately, these conditions are quite reproducible from heat to heat.

The shape of the solidification front in the ingot is a good indicator of the freezing conditions and usually is described by giving the depth of the metal pool. Generally, a shallow pool provides the best metallurgical conditions and results in a more vertical freezing pattern and better soundness. A shallow pool is favored by using a high electrode/ingot fill ratio, high voltage, low current, and a deep slag cap. The latter three factors are related to the slag composition selected. Low melt rates tend to give improved slag-metal refining effects and a better ingot cleanliness. With regard to productivity, however, a high melt rate is desirable; this means that often a compromise is made between the low current to give best ingot properties and the high current to provide more economic melting (Machner, 1973). The success of the ESR process relies heavily on selecting the correct slag composition for the wide variety of high-alloy steels.

Typical slag compositions generally used for the remelting of high-alloy steels are listed in Table 21 (Schlatter, 1972). To assure a high level of consistency in the preparation of these flux mixes, proper quality control monitoring is advisable.

TABLE 21 Typical Flux Compositions for Remelting Alloy Tool Steels.

Composition (percent)						
CaF_2	CaO	Al_2O_3	MgO	SiO_2	TiO_2	Application
70	--	30	--	--	--	General purpose flux.
70	15	15	--	--	--	For high-speed steels; good desulfurization.
55	15	25	3	2	--	For hot-work die steels.
40	--	50	10	--	--	Hopkins flux for die steels.
40	30	30	--	--	--	Low-melting point fluxes for
30	30	30	--	10	--	high-carbon tool steels.
30	--	40	10	20	--	Hopkins flux for tool steels.
50	--	25	--	--	25	For sulfurized tool steels.
50	--	25	--	25	--	
35	15	--	5	10	35	Conductive, for cold start-up.

SOURCE: Schlatter, 1972.

Usually, the ESR fluxes are made up of relatively high-purity raw materials to minimize compositional variations in the remelted steel and to avoid undesirable impurity pickup. The following maximum impurity levels, in weight percent, should not be exceeded in prefused or premixed fluxes of the CaF_2-CaO-Al_2O_3 type used for solid slag starting:

Impurity	Weight Percent
C	0.03
P	0.005
S	0.02
PbO	0.003
Bi_2O_3	0.002
SiO_2	0.50
FeO	0.20
MnO	0.20
MgO	0.50
$Na_2O + K_2O$	0.20

2. Technical Advantages

Many advantages have been claimed for the ESR process since its industrial introduction, but its shortcomings and limitations also must be considered if it is to be applied properly. Nevertheless, the ESR process does provide features not available in other secondary remelting/refining processes. The important advantages of ESR for stainless and high-alloy steels over conventional air-melting and ingot casting are listed below:

1) Process flexibility and versatility*
   a) Electrode/ingot geometry can vary widely
   b) Relatively easy shape manufacture
   c) Electrode change can be employed (within limits)
   d) Composite electrodes, composite ingots
2) Control of solidification rate and direction
   a) Macrosegregations minimized
   b) Microsegregations reduced
   c) Better distribution of secondary phases
   d) General structural uniformity much improved
3) Yield improvement considerations
   a) Increased ingot weights (with complex alloys)
   b) Reduced hot top losses
   c) Smooth ingot surfaces
   d) Improved hot workability
4) Macro- and micro-cleanliness
   a) Nonmetallic inclusions minimized
   b) Highly isotropic properties
   c) Better fatigue properties
   d) Good cold forming characteristics
   e) Good polishability
5) Composition control*
   a) Excellent desulfurization (or sulfur retention)
   b) Retention of volatile alloying elements
   c) Reduction of combined oxygen content
   d) Minor alloy corrections possible

The many advantages have been attributed to the large number of degrees of freedom that the ESR process provides (Duckworth and Hoyle, 1969). The increasing requirements placed on modern products demand manufacturing processes providing flexibility and versatility in the selection of process parameters. However, many degrees of freedom greatly increase the process complexity and require a careful balance of compromises. The increased process complexity also leads to an array of different and unique control problems (e.g., maintenance of desired flux chemistry, consistent slag pool depth and temperature, composition uniformity in ingots, and atmosphere control).

---

* Compared with the VAR process.

The excellent flexibility and versatility of the ESR process is reflected well by the many different furnace designs and the various operating modes as well as the possibility of engineering a remelting installation for a very specific purpose (shapes). Freedom in electrode/ingot geometry, shape manufacture, use of electrode change, and the making of composite ingots are factors of considerable importance in the production of particular high-alloy steel products.

Control of solidification conditions and the resulting ingot macro- and micro-structure are of prime concern in many ESR applications, particularly for high-alloy tool steels with their complex structures. Macrosegregations (e.g., freckles), radial segregations, and tree-ring patterns generally are minimized but certain typical defects, unique to consumable electrode remelting processes, also can be encountered. A reduction in microsegregation and improvement in the distribution of secondary phases usually are attained. The capability of producing larger and heavier ingots with good structural uniformity is often a decisive factor in using ESR for high-alloy steel products of large cross section.

A major claim of the ESR process – often debated extensively by experts – is yield improvement as the result of several factors cited above. As will be discussed below, the yield improvement must pay for the added cost of the remelting process and its concomitant auxiliary operations.

The hot workability of high-alloy steels, as measured by hot torsion or tensile tests, generally is improved by ESR. Figure 7 shows the results of tests performed on M-2 high-speed steel. The hot ductility improvements are usually modest but may contribute to better yields (Lowe and Hogg, 1973). Macro- and micro-cleanliness are improved substantially over air-melted material. Mechanical properties in the transverse direction and fatigue properties in general benefit greatly from effective removal of inclusions and good dispersion of the remaining very small globular inclusions (mostly oxides, rarely sulfides). The most important aspects in composition control are the excellent desulfurization potential and retention of volatile alloying elements. Retention of manganese and nitrogen in special stainless steels is of particular benefit. The potential of introducing nitrogen by remelting under a positive nitrogen pressure has been demonstrated successfully for austenitic stainless steels (Kubisch and Holzgruber, 1971). The reduction of combined oxygen content (in oxides) is due to the removal of inclusions, which depends strongly on the type of alloy and selected process parameters. The possibility of making minor alloy adjustments through the use of special slags or carefully metered additions is sometimes considered useful. The loss of easily oxidizable elements generally is compensated for in the electrode or by making alloy additions during remelting.

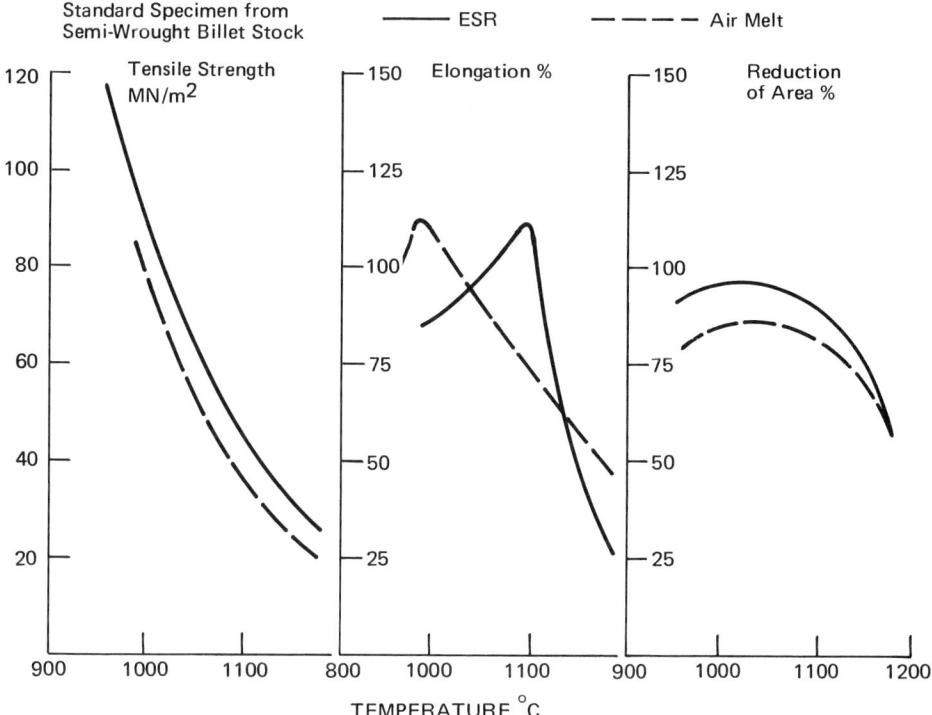

FIGURE 7   Hot Tensile Properties of Air-Melted and ESR M-2 High-Speed Steel (from Dewsnap and Schlatter, 1974).

3. Process Limitations

The disadvantages and limitations of ESR as compared to conventional air melting and casting may be summarized as follows:
1) Considerable process complexity
2) Additional (thermal) handling of electrodes
3) Poor power efficiency
4) High remelting cost (fluxes, power)
5) No hydrogen removal (extra degassing required)
6) No removal of volatile tramp elements
7) Close composition control often difficult (reactive elements)
8) Fluxes are potential source of undesirable residual elements
9) Carbide structure refinement marginal (tool steels)

Some of the above disadvantages apply to specific alloy groups only. In contrast to VAR, no removal of hydrogen or volatile tramp elements occurs during ESR. As discussed above, process complexity has two sides – the advantage of excellent flexibility and versatility and the disadvantage of creating some unique and difficult control problems. Modern ESR installations for the production of premium-quality high-alloy steels are relatively complex. They must be run by well-trained operators and supervised by technically competent personnel. The necessary production steps in manufacturing electrodes and their proper treatment and preparation for remelting are primarily of economic concern, as are the poor power efficiency and high flux costs. Other limitations deal with quality aspects of the materials melted by ESR (e.g., lack of hydrogen removal) that may require a degassing treatment of the primary melt and a closed remelting system with a controlled atmosphere. Hydrogen problems are of predominant concern in low-alloy steels and ingot sizes larger than those commonly produced in this material group. For optimum solidification and structure control, ESR (like VAR) does not provide the ultimate system. A heat source for independent adjustment of melt rate and thermal gradients within feasible limits would be preferred for this purpose, especially for structure-sensitive alloys.

Composition control and uniformity within narrow limits throughout an ingot is difficult to achieve with reactive alloying elements and minute additions of beneficial elements in parts per million unless extensive precautionary steps are taken during electroslag remelting. Making additions through specific slag components or continuous feeding of alloys to the molten pool often causes inconsistency, non-reproducibility, and even ingot defects (unmelted particles).

4. Shapes

In considering the shapes that could provide potential cost-effective advantages in the manufacture of semi-finished stock or parts of these steels, the following tabulation serves as a guide:
1) Stainless steels – composite round and square ingots; slabs for direct plate rolling; composite slabs for clad plate; hollow ingots for direct extrusion; and valve, vessel and pump components for chemical and nuclear applications.
2) High-strength steels – slabs for direct plate rolling; armor plate (composite plates); and hollow ingots for pressure vessels, missile components, and gun tubes.
3) High-alloy tool steels – square ingots for direct rolling; rectangular and slab shapes for strip and sheet; rolls with journals cast-on; composite rolls; round ingots for roll discs and slitter knives; and composite ingots for large cutters.

Shaped ingots have been produced in stainless steels and specialty alloys but little work has been done on a larger scale in regard to high-alloy high-strength and tool steels. Some special austenitic stainless steels are being produced to some extent by ESR in slab ingots or rectangular shapes in

Sweden and in the USSR where the Soviet bifilar setup is preferred. Round-corner square ingots and rectangular shapes of an aspect ratio of approximately 2 to 3 are in limited production in high-alloy tool steel, particularly in Europe. Shaped ingots are used for these alloys mainly to produce billets and bars of large cross section with good structural uniformity. The more favorable solidification conditions and reduced macrosegregation tendency in these ingots are most advantageous for these products. Square ingots often are preferred for direct roll cogging of tool steels. The most desirable practice for these steel grades (and some other alloys) would be the economical production of relatively small cross-section ingots for direct secondary rolling to bar and coil. Direct application of as-cast ingots is growing because of some unique properties of the cast structure (isotropy, wear characteristics). Hollow ingots have not been produced industrially in alloys of these three groups.

5. Conclusions and Recommendations

   a. Conclusions

   (1) Electroslag remelting has found significant applications in the production of premium-quality high-alloy steels. Special stainless steels, a number of specialty alloys, and high-alloy tool steels benefit the most from ESR processing in regard to specific quality and technological improvements, processing advantages, and economics.

   (2) The viability of ESR for many high-alloy steels compared to conventional air melting relies primarily upon balancing the remelting costs against the realization of significant metallurgical, technological, and yield benefits.

   (3) ESR is a valuable alternative process to VAR, providing features and advantages not available with the VAR process (e.g., desulfurization, retention of volatile alloying elements, wide process and technological flexibility).

   (4) The capability of ESR to produce high-integrity shaped ingots and castings of simple configuration is a feature of increasing importance for high-alloy steel products.

   (5) Principal deterrents against wider application of ESR for the large variety of high-alloy steels are high operating costs (fluxes and power consumption) and limitations in metallurgical and process control (e.g., lack of hydrogen removal, structure refinements, consistent control of reactive alloying elements).

b. <u>Recommendations</u>

   (1) Develop a production process to cast several small ingots by electroslag melting of one large electrode. Small cross-section ingots (4 to 10 inches in diameter) would provide important economical and metallurgical incentives for the manufacture of a large variety of bar and wire products of many high-alloy steels.

   (2) Evaluate thoroughly the as-cast properties of ESR high-alloy steels to further promote the direct application of shaped ingots and castings.

   (3) Develop techniques and procedures to maintain consistent remelting parameters for the utilization of maximum refining capability and most economical melting conditions throughout the melting process. Improved control over slag pool depth and slag composition is needed and continuous measurement of slag bath temperature is desirable to control the remelting process.

   (4) Improve solidification control of complex structure-sensitive high-alloy grades to produce ingots of larger cross section with a high degree of structural refinement and consistency.

   (5) Develop slags with low CaF_2 content (< 40 percent CaF_2) and the recycling of spent slag to counteract the growing shortage of high-grade CaF_2 and its rapidly increasing cost.

   (6) Study the thermodynamic and physical properties of principal ternary and quaternary slag systems important to ESR thereby facilitating the selection of slags for remelting of alloy steels and specialty alloys.

   (7) Investigate the effective use of electrochemical reactions occurring in DC melting and the effect of low-cycle AC and AC rectification on such reactions.

D. Electroslag Technology for Other Metals and Alloys

Considerable experimental work has been conducted during the past two decades in the United States and abroad (most notably in the USSR) on electroslag remelting of reactive, refractory, and heavy metals and alloys. Metals and alloys in these categories that have been electroslag melted include:

1) Reactive metals and alloys — titanium, Ti-8Mn, Ti-6Al-4V, zirconium, thorium, U-7.5Nb-2.5Zr, aluminum, Al-2.5Mg, and chromium.

2) Refractory metals and alloys — molybdenum.

3) Heavy metals and alloys — copper, monel, chromium bronze, and tin-phosphorus bronze.

1. Process Parameters

The principles of electroslag melting and a basic description of the process have been summarized in section IV. In all of the work reported in this section, stationary water-cooled copper crucibles of varying dimensions have been used to contain the ingot. Single-phase AC was used as the power mode in the USSR as well as for the Mellon Institute work on electroslag melting Ti-6Al-4V (Bhat, 1971). In all of the remaining electroslag research on titanium, zirconium, uranium, aluminum, molybdenum, and their alloys, straight polarity DC power was employed, primarily because these melts were conducted in modified vacuum arc furnaces. Electrode material for electroslag remelting reactive, refractory, and heavy metals and alloys can range from compacted sponge (Ausmus and Beall, 1968; Nafziger and Calvert, 1971) to rolled bar stock (Bhat, 1971), previously melted ingots (Nafziger and Calvert, 1971), and powder metal bar (Calvert et al., 1971). The electrode stock is generally the most important input parameter in determining ingot impurity levels. Electroslag remelting of most reactive, refractory, and heavy metals requires an inert furnace atmosphere inasmuch as the slag is an inadequate protective covering against atmospheric contamination and unwanted oxidation. Usually a backfilled atmosphere of argon (Morozov et al., 1962; Gurevich et al., 1963) or helium (Ausmus and Beall, 1968; Nafziger and Calvert, 1971; Calvert et al., 1971) is used as the furnace atmosphere for electroslag remelting titanium, zirconium, and molybdenum and their alloys. A positive pressure of argon also has been used to electroslag remelt Ti-6Al-4V (Bhat, 1971).

Electroslag remelting of reactive, refractory, and heavy metals places great restrictions on suitable slag compositions. For reactive metals, slags containing oxides are inappropriate because oxygen readily transfers from the slag to the molten metal at high temperatures. Since the liquidus temperature of the slag should be near that of the metal or alloy for satisfactory electroslag remelting, alkaline earth and rare earth fluoride (including scandium, yttrium and lanthanum) represent the only suitable slag compositions for reactive metals. Other halides possess unsuitably low liquidus temperatures, high vapor pressures, and high temperature instability, although some can be used to electroslag remelt aluminum and its alloys. A variety of prefused fluoride slags has been used to electroslag remelt titanium, including CaF_2, BaF_2, MgF_2, MgF_2-LaF_3, CaF_2-LaF_3, CaF_2-MgF_2, CaF_2-SrF_2, and CaF_2-Y (or Gd) (Nafziger, 1969; Nafziger and Riazance, 1972). CaF_2 has provided the most satisfactory slag composition for titanium from the standpoint of economics, availability, and ingot quality. For electroslag remelting zirconium, CaF_2 and LaF_3 slags have been used (Nafziger and Calvert, 1971). Slags composed of CaF_2, CaF_2-MgF_2, and CaF_2-Ca were employed to electroslag remelt uranium alloys (Cadden et al., 1973). In general, cryolite-alkali chloride, chloride-fluoride, alkali fluoride-AlF_3, and ternary Li-Na-K cryolite slags have been used to melt aluminum and its alloys (Andreev et al., 1967; Chulkov et al., 1970; Wilson et al., 1969). In all cases, acid grade fluorspar (containing a minimum of $\sim$ 98 percent CaF_2) is required to electroslag remelt reactive, refractory, and heavy metals. It is especially important to remove all traces of moisture and undesirable volatiles from the slag prior to melting, and this customarily is accomplished by fusing the slag in an inert atmosphere. Other impurities are less important but optimum levels in weight percent are: Al, 0.02 to 0.06*; C, 0.01 to 0.10*; CO_2, 0.01 to 0.04*; Fe, 0.02; Mg, 0.05; Si, 0.006 to 0.005*.

Slags with very high liquidus temperatures are required for electroslag remelting refractory metals and alloys. Prefused Y_2O_3 with or without C and/or Y additions has been satisfacroty for electroslag remelting molybdenum (Calvert et al., 1971). Soviet investigators have used CaF_2-$BaCl_2$ with CaC_2, and CaF_2-NaCl or CaF_2-NaF slags to electroslag remelt copper alloys (Latash and Medovar, 1966; Vainshtok et al., 1971; Korenyuk and Didkovskii, 1960).

2. Technical Advantages

    a. Ingot Quality

    A major advantage offered by electroslag remelting reactive metals over more conventional secondary melting processes (e.g., VAR) is the smooth, directly workable ingot surface that results from the freezing of a thin layer of slag between the crucible and ingot metal. Therefore, expensive

---

* Second number is the maximum value.

machining usually is not required for ESR ingots prior to working. In addition, the characteristic ingot crown found on VAR ingots that requires scalping is not present in ESR ingots. A comparison of the surfaces of ESR and VAR molybdenum ingots is shown in Figure 8. Increased yields also have been noted in

(a)   (b)

FIGURE 8 ESR (a) and VAR (b) Molybdenum Ingots (from Calvert et al., 1971).

ESR material (Morozov et al., 1961; Gurevich et al., 1963; Bhat, 1971). Axial grain orientation due to progressive solidification and control of melting rates is accomplished more easily during electroslag melting of reactive, refractory, and heavy metals in comparison with other conventional melting processes. Differences in macrostructures of ESR and arc-cast molybdenum ingots are shown in Figure 9. Also, the ESR process often results in a reduction in size and a uniform dispersal of nonmetallic inclusions in the ingot metal (Cadden et al., 1973). This may be one of the principal advantages in electroslag remelting uranium alloys. Casting flaws were noted in arc-melted material that were attributed to the agglomeration of oxide particles.

With respect to segregation, niobium tends to concentrate at the center of electroslag remelted U-7.5Nb-2.5Zr ingots. However, this was offset by a significant reduction in banding segregation (Cadden et al., 1973). Based on earlier work conducted at the Carnegie-Mellon Institute, the Sandia Corporation reportedly is electroslag remelting depleted uranium into 6 to 8 inches-diameter ingots. The most satisfactory uniformity in ingots was achieved by the electroslag process.

In general, alloying element retention is acceptable in electroslag remelted Ti-6Al-4V (Beall et al., 1967; Bhat, 1971) and U-7.5Nb-2.5Zr (Cadden, 1973). In some cases, oxygen can be removed during electroslag melting. This occurs when U-7.5Nb-2.5Zr was melted with a CaF_2-Ca slag, when molybdenum was electroslag melted with a Y_2O_3-C and/or Y slag resulting in improved ductility in the wrought products, and when copper was electroslag remelted with a fluoride slag or by adding graphite and CaC_2 to a $80CaF_2$-$20BaCl_2$ slag (Pogodin-Alekseev and Syrovatkin, 1964). Reduction in hydrogen has been reported in aluminum alloys when $90(KCl + MgCl_2)$-$10MgF_2$ or $47KCl$-$30NaCl$-$23Na_3AlF_6$ slags were used in the ESR process (Chulkov et al., 1970). Although electroslag remelting of titanium with fluoride slags increased the fluorine content of the ingot metal, no significant effect on corrosion-resistance properties, as compared with VAR material, was observed (Armantrout and Nafziger, 1969).

Chromium bronzes have been electroslag remelted in an argon atmosphere with CaF_2-(2.5-10) NaCl slags (Vainshtok et al., 1971). No defects or inclusions were noted, and the material exhibited improved hot workability and weldability.

Although earlier efforts to electroslag remelt chromium using $CaF_2Al_2O_3$ slags resulted in essentially no nitrogen and oxygen removal, recent unpublished work (Mitchell, 1974) indicates a 56 percent decrease in nitrogen (from 0.25 to 0.11 percent) and a 98 percent decrease in oxygen (from 0.50 to 0.01 percent) from electrode to ingot. These melts were conducted using CaF_2-CaO slags in an argon atmosphere. Excellent ingot structures and surfaces were obtained. These results demonstrate that chromium or high-chromium alloys (e.g., 80 percent chromium ferrochromium) can be electroslag remelted successfully. Seventy percent ferrochrome containing 0.22 to 0.33 percent carbon has been electroslag remelted successfully into 4-inch-diameter ingots using single-phase AC at 900 to 1000 A and 30 to 35^{-1} V

FIGURE 9 Comparative Macrostructures of ESR (a) and VAR (b) Molybdenum Ingots (from Calvert et al., 1971.)

(Kammel and Winterhager, 1968). Maximum melting rates of 61.6 lb/hr^{-1} and an energy consumption of approximately 1.2 kWhr kg^{-1} were obtained. The slags were reported to be CaF_2-based with additions of Al_2O_3, CaO, and Fe_2O_3. Also, copper and aluminum have been electroslag remelted. In the first case, oxygen removal was accomplished readily by using NaF slags containing 30 percent by weight or more CaF_2. For aluminum, slags in the system $KCl-NaCl-Na_3AlF_6$ containing 10 percent cryolite were used to obtain highly refined (i.e., low Al_2O_3) metal by the electroslag process.

Soviet investigators have electroslag remelted 6-inch-diameter monel ingots at 1800 to 2500 amps and 46 volts using electrodes 2.8 to 3 inches in diameter and a dried $20CaF_2-25CaO-15MgO-40Al_2O_3$ slag. The manganese content was found to decrease 7 to 10 percent from electrode to ingot, whereas other alloying components remained unchanged. The major advantages were the lack of shrinkage porosity and hairline cracks in the ESR ingots compared with ingots produced by conventional casting methods (Lekarenko et al., 1964). The tendency to shrinkage porosity in ingots is due to the wide crystallization range of monel metal. Hairline cracks cannot be tolerated in monel rods that are used in the aircraft industry. Thus, the ESR process may provide a good method for producing extruded tubes from more forgeable ingots.

Thorium has been electroslag remelted by the Battelle-Columbus Laboratories (Hoffman, 1974). In addition, the Energy Research and Development Administration Ames Laboratory at Iowa State University is electroslag remelting thorium with oxide slags in an attempt to transfer the radioactive products into the slag. Subsequent electron beam melting to remove the oxides is envisioned (Schmidt, 1974).

At present, the Pitman-Dunn Laboratory of the Frankford Arsenal is experimenting to assess the feasibility of electroslag remelting high-strength aluminum alloys (such as 7075). The objective is to obtain ingots with fewer impurities, improved homogeneity, and better grain size and orientation than those obtained in conventional casting operations. Fluxes composed of 80 percent by weight cryolite and 20 percent KCl show promise; however, persistent magnesium losses during melting have been a problem. Dendrite arm spacing is generally less than that reported for conventionally cast ingots (Markus, 1974).

b. Mechanical Properties

Soviet investigators first demonstrated that electroslag melting generally improves the mechanical properties of Ti-8Mn ingots over those that were double vacuum arc melted. Additions of aluminum (1.25 to 2.50 percent) and manganese (0.8 to 1.5 percent) increase the tensile strength of as-cast electroslag remelted material (Gurevich et al., 1963). When molybdenum was electroslag remelted into 3.6-inch-diameter ingots with smooth surfaces and axial grain orientation, upset-forging could be accomplished at temperatures as low as 1112° F with 1292 to 1382° F considered optimum (Calvert et al., 1971). Ductile-to-brittle transition temperature (DBTT) comparisons showed that stress-relieved sheet prepared from ESR molybdenum should fabricate more readily by bending than that prepared from VAR ingots. All DBTT values for ESR sheet were below room temperature, indicating suitability for room-temperature fabrication, while the same parameter for VAR material was well above room temperature (Figure 10). Tensile strength of sheet stress-relieved from room temperature through 1112° F is independent of carbon, oxygen, or yttrium impurities. At such temperatures, sheet from the ESR ingots possesses a higher tensile strength (by 18,000 to 20,000 psi) than does sheet from VAR ingots. At 1832° F, a rapid decrease in tensile strength of sheet from ESR ingots with high yttrium levels is noted. At 2532° F, the sheet with high oxygen content is twice as strong as the VAR or ESR materials with high yttrium and carbon contents. At temperatures up to 3632° F, tensile

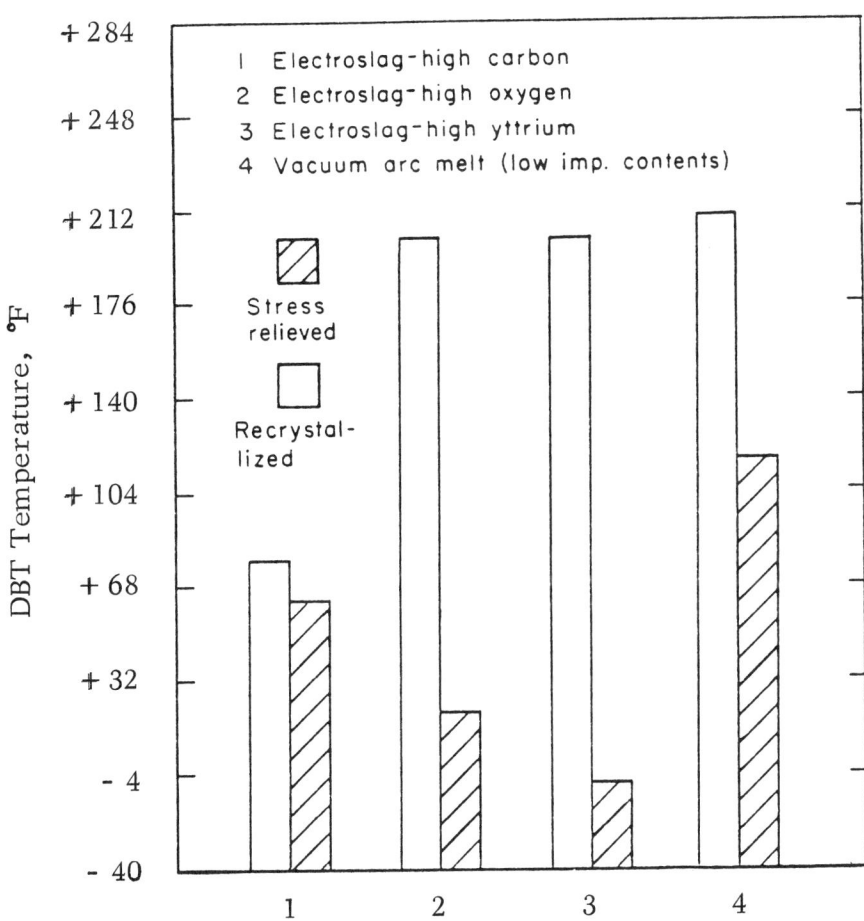

FIGURE 10 Ductile-to-Brittle Transition Temperature of 0.15 cm Sheet Rolled from Electroslag and Arc-Melted Molybdenum Ingots (from Calvert et al., 1971).

strengths for all ingots (VAR and ESR) were nearly equal. Up to 1112° F, the yield strengths follow the trends of the tensile strengths. At 1832° F, the yield strengths of arc-cast sheet and high carbon ESR sheet exceed those of the high oxygen and yttrium ESR sheets by 10,000 to 19,000 psi. Preliminary data indicate that the 100-hour stress rupture strength of material derived from ESR ingots is 16,000 psi higher than the averages for commercially pure molybdenum. Thus, electroslag remelting of molybdenum results in a minimum loss of metal by oxidation of the ingot during fabrication due to low forging temperatures. Superior mechanical properties for ESR material that may be due to compositional variations between ESR and VAR ingots also have been observed (Calvert et al., 1971). It also is possible that the size reduction and the distribution of inclusions in ESR zirconium ingots could have a beneficial effect on mechanical properties.

In many cases, mechanical properties of wrought material derived from ESR reactive and heavy metals and alloys are comparable to those of VAR material. In the case of titanium, tensile properties are usually comparable (Figure 11). Tensile and yield strengths of specimens cut from 0.06-inch-thick sheet derived from forging and rolling of zirconium ingots melted from reactor-grade sponge were comparable for ESR and VAR material (Nafziger and Calvert, 1971). Mechanical properties of ESR copper ingots also were comparable to those of conventionally cast ingots (Pogodin-Alekseev and Syrovatkin, 1964).

Coupons cut from 0.06-inch-thick zirconium sheet derived from ESR material showed comparable corrosion resistance to steam and saturated steam, ranging from 680 to 750° F and 2,705 and 1,500 psi, respectively, as that derived from VAR material. Zirconium electroslag melted from high-iron sponge shows less corrosion resistance than that melted from reactor-grade sponge, regardless of melting mode. Material electroslag melted with CaF_2 appears to possess greater corrosion resistance than that melted with LaF_3 for both sponge sources (Nafziger and Calvert, 1971).

3. Process Limitations

    a. Ingot Quality

One of the major disadvantages of the ESR process when used to melt reactive metals is the observable lack of refining capability. When leached and dried titanium sponge is used as the electrode material, hydrogen, oxygen, nitrogen, sponge reductant (such as magnesium), and volatile species levels remain essentially the same in the ingot as in the electrode. Either vacuum distilled or electrolytic sponge is required to maintain low levels of the above impurities in ESR ingots (Ausmus and Beall, 1968; Armantrout and

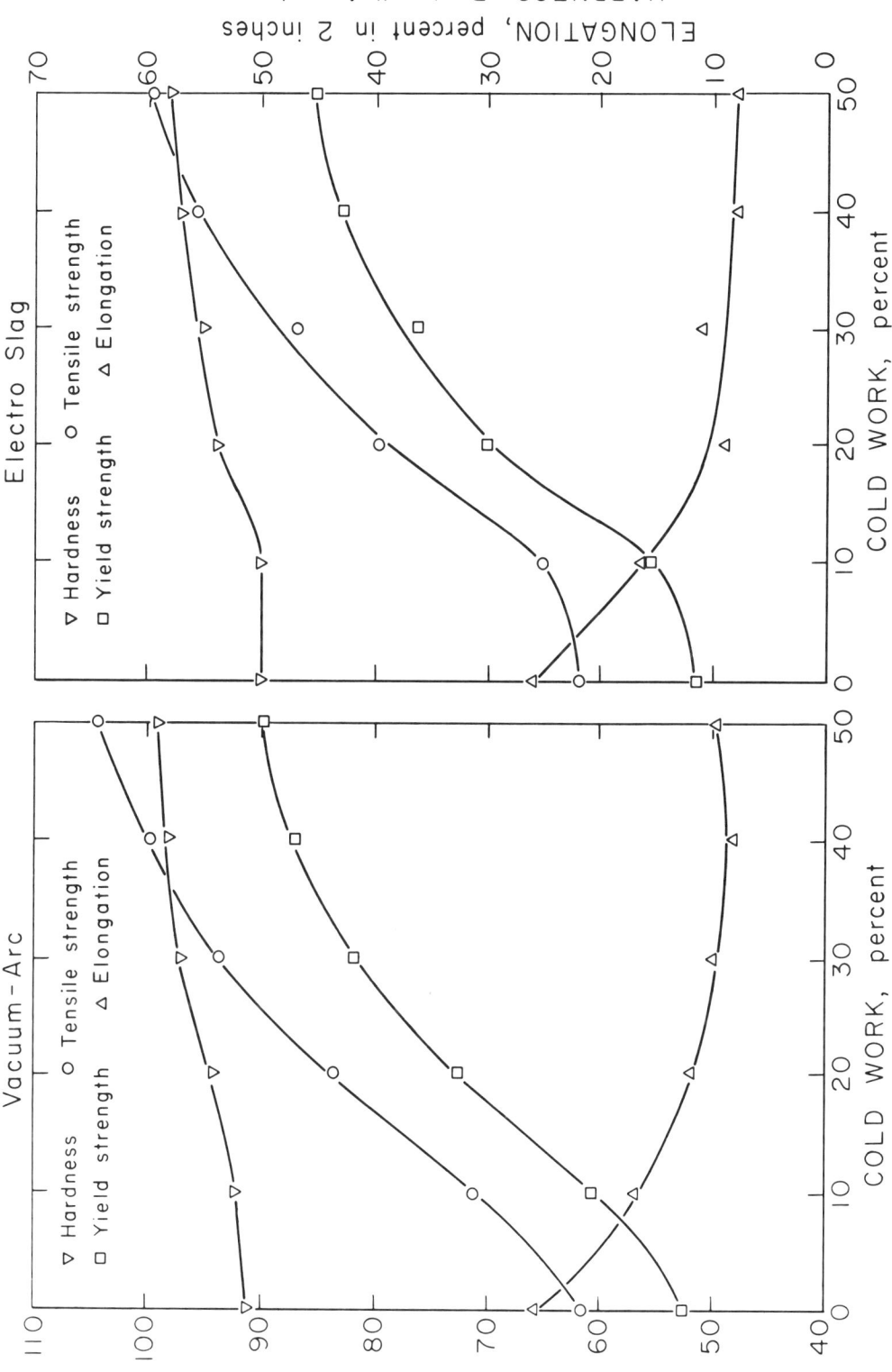

FIGURE 11 Tensile Properties and Hardness of 0.16-cm-Thick Titanium Sheet Rolled from Vacuum-Arc Melted Material and Electroslag Melted Material (from Armantrout et al., 1970).

Nafziger, 1969; Bhat, 1971). These observations apply not only to small-scale (4 inch diameter) ingots but also to slab-shaped ESR titanium ingots up to 8 by 32 by 28 inches (Morozov et al., 1961; Beall et al., 1969). Electroslag remelting of zirconium sponge and ingot material also resulted in minimal refinement. Electrode stock is the most significant tested parameter for determining eventual ingot quality, whereas melting rate, DC polarity, and furnace atmosphere play minor roles (Nafziger and Calvert, 1971). In the case of the U-7.5Nb-2.5Zr alloy, ingot oxygen contents were not reduced by electroslag remelting with CaF_2 slags. Some uranium and zirconium was lost to the slag and deposited as particles which caused a reddish color. Uranium alloy ingots electroslag remelted with a CaF_2-1.0 weight percent MgF_2 flux were significantly lower in oxygen than those melted with a CaF_2 slag, although ingot surfaces were much rougher. Melts conducted with CaF_2-(2.2-4.3) weight percent calcium slags resulted in ingot oxygen levels comparable to those in ingots which were vacuum arc remelted, as well as rough ingot surfaces.

With respect to electroslag remelting aluminum alloys, two problems are evident in the limited studies that have been conducted: (1) a relatively high current density is required to maintain uniform melting due to high heat losses, and (2) the cooling water tends to cool the crucible too rapidly in DC melting Al-2½Mg alloys (Wilson et al., 1969). Poor ingot sidewalls and excessive magnesium losses limit the utility of electroslag remelting this alloy. In addition, aluminum and aluminum alloys require fluxes with low liquidus temperatures which, in turn, implies the presence of alkali compounds that are difficult to handle in the electroslag process.

In general, the microstructures of samples taken from ESR reactive metal ingots show very fine and randomly distributed inclusions. This was particularly evident for zirconium, wherein all as-cast ingots exhibited a basket-weave structure at 250 magnification. The structure showed evidence of inclusions, possibly containing oxygen, nitrogen, carbon, and/or one or more transition metal elements. Wrought material from ingots melted with a CaF_2 slag contained randomly distributed CaF_2 particles. No such particles were observed in material from ingots melted with a LaF_3 slag (Nafziger and Calvert, 1971). Fluoride inclusions would be detrimental to zirconium since it is used primarily in water-moderated nuclear reactors.

b. <u>Mechanical Properties</u>

When reactive metals are electroslag melted, the most significant degradation occurs in impact strengths of wrought material as compared with vacuum arc melted material. This is demonstrated in Table 22 for titanium (Morozov et al., 1961; Gurevich et al., 1963; Armantrout and Nafziger, 1969), Ti-8Mn (Gurevich et al., 1963), and zirconium (Nafziger and Calvert, 1971). In the case of titanium, vacuum distilled sponge yielded material with the best impact strength, with leached and dried material significantly worse (Armantrout and Nafziger, 1969).

TABLE 22 Average Impact Strengths of Plate Fabricated from Electroslag Melted and Vacuum-Arc Melted Titanium and Zirconium Ingots.

Material	Electroslag Melted (foot pounds)	Vacuum-Arc Melted (foot pounds)
Titanium, Mg reduced, leached and dried sponge	16	--
Titanium, vacuum-distilled sponge	86	171
Titanium, Mg reduced, gas swept and/or water-leached sponge	23	--
Titanium, Na reduced, leached and dried sponge	3	--
Zirconium, high-iron sponge	7.5 ± 1.5	9.7 ± 2.4
Zirconium, reactor-grade sponge	6.2 ± 0.7	18.2 ± 1.2

SOURCE: Armantrout and Nafziger, 1969; Nafziger and Calvert, 1971.

The source of electrode material also determines to a great extent the comparable mechanical properties of electroslag melted zirconium. In the case of material melted from high-iron sponge, lower tensile and yield strengths were noted for material melted with CaF_2, and especially LaF_3, compared to those for vacuum-arc melted material. However, ductility as a function of melting mode, sponge stock, and flux composition for electroslag melted materials was unaffected (Nafziger and Calvert, 1971).

For titanium, room-temperature minimum bend radius was determined for sheet from 4.1-inch-diameter ingots in the as-hot-rolled condition and in sheet reduced 50 percent by cold rolling. The bending characteristic of VAR titanium appeared superior to that of electroslag melted material, especially for hot-rolled sheet, although this difference in bend ductility may have been due to prior ingot preparation. The vacuum-arc melted ingots were machined, whereas this was not the case for the electroslag melted material (Armantrout et al., 1970).

4. Economic Aspects

It is extremely difficult to estimate the relative cost of electroslag and vacuum-arc melting reactive, refractory, and heavy metals inasmuch as production experience is nonexistent for these metals and alloys. In any case, comparative costs are difficult to define because of the wide variation in labor costs, overhead, capital costs, and depreciation. Perhaps the greatest single cost item in all processes using a consumable electrode is the electrode material and its fabrication. Cost factors to be considered when evaluating secondary melting processes include: capital, labor, overhead, molds, electrode production, depreciation, cooling water, power, and if applicable, slag. In most cases, the costs for each factor in VAR and ESR will be approximately equal, with the exception of the last four items. Although ESR power and slag costs are greater than those for more conventional techniques, the fact that double and triple arc melting often is required to obtain acceptable ingots and the aforementioned advantages of ESR reactive, refractory, and heavy metals and alloys may offset at least partially its initial higher cost. Costs are expected to be comparable for the two processes.

5. Energy Requirements

Available data on energy required to electroslag remelt reactive, refractory, and heavy metals on an experimental basis are presented in Table 23. While ESR requires more energy than VAR, the same arguments apply in this case as for increased costs inherent in electroslag remelting.

6. Shapes

Slab-shaped titanium and titanium alloy ingots have been electroslag melted successfully in the past. Rectangular-shaped titanium ingots 3.7 by 7.0 inches at the top tapered to 3.0 by 6.5 inches at the bottom and 10.6 inches in length were prepared using a conventional electroslag furnace with an appropriately shaped water-cooled copper crucible, DC power, and a CaF_2 flux (Ausmus and Beall, 1968). Such ingots were readily side-forged and rolled to sheet with some localized grinding required. Tensile data were comparable to those of sheet fabricated from vacuum arc melted material. The success of this program prompted the design and construction of a furnace capable of electroslag melting ingots 3 by 20 inches in cross section and 22 inches in length in a water-cooled copper crucible. The titanium sponge electrode was 4 by 16 inches in cross section. Sound ingots were prepared but were not suitable for direct conversion to sheet or plate without prior machining. Some machining of the outer surface would be required. Subsurface porosity was minimal, in contrast to double vacuum-arc melted material. Higher hardnesses at the bottom of the ingot were attributed to initial flux impurities which were gettered by the molten titanium (Beall et al., 1969).

TABLE 23 Average Values of Melting Rates and Energy Requirements for Melting Reactive, Refractory, and Heavy Metals by Various Processes.

Commodity	Melting Process*	Melting Rate (2.2 lb min^{-1})	Power (kW)	Energy Consumption (kWhr 2.2 lb^{-1})	Reference
Titanium	VAR 4 in. ⌀	1.36	110.2	1.32	Armantrout and Nafziger, 1969
	7.2 by 20.4 in.	4.70		0.92	Beall et al., 1969
	ES 4 in. ⌀	0.91	94.5	1.54	Armantrout and Nafziger, 1969
	7.2 by 20.4 in.	0.91-2.72	432.0	1.76-4.41	Beall et al., 1969
Zirconium	VAR 8 in. ⌀	2.04	75.0	0.63	Beall et al., 1968
	ES 4 in. ⌀ (LaF$_3$)	3.45	151.9	0.88	Nafziger and Calvert, 1971
Uranium Alloys	VAR	No data	No data	No data	
	Ind. 5.2 in ⌀	5.87	191.7	0.94	
	7.2 in ⌀	7.28	130.0	0.28	Cadden et al., 1973
	ES 5.2 in. ⌀		85.2		
	7.2 in. ⌀	No data	144.0	No data	Cadden et al., 1973
Molybdenum	VAR 3.6 in. ⌀	No data	99.9	No data	
	ES 3.6 in. ⌀	0.39	144.1	6.1	Calvert et al., 1971
Copper	VAR 3.2 in. ⌀	No data	68.8	No data	Beall et al., 1968
	ES 3.2 in. ⌀	0.89	34.1	0.74	Nafziger, unpublished

* VAR = vacuum-arc first melt; ES = electroslag melting; Ind. = induction melting.

Electroslag remelting of Ti-6Al-4V into 2.5 by 4 inches slab-shaped ingots has been conducted (Bhat, 1971). Rolled bar stock served as the electrode material and dried argon was used to maintain a positive pressure during these AC melting experiments. In addition, titanium sponge was pressed isostatically to form 10 by 19 by 80 inches long electrodes to prepare 12 by 36 by 30 inches long ingots. This work was generally unsuccessful (Bhat, 1971).

7. Conclusions and Recommendations

   a. Conclusions

   (1) At present, there are no known commercial electroslag operations for melting reactive, refractory, or heavy metals and alloys in the United States or, probably, abroad. Commercial acceptance would be likely only if large ingots having a variety of shapes could be produced experimentally possessing the specific advantages inherent in electroslag melting small ingots.

   (2) The considerable experimental work that has been conducted on electroslag remelting reactive, refractory, and heavy metals and alloys has shown that satisfactory ingots up to 4 inches in diameter with smooth, directly workable external surfaces and axial grain orientation can be obtained from a variety of electrode materials using an inert furnace atmosphere.

   (3) Greater yields have been attained in ESR material, and inclusions are often dispersed and reduced in size. Although electroslag melted titanium and zirconium contain fluoride inclusions that are detrimental to zirconium used in nuclear applications, the corrosion resistance of electroslag melted titanium and zirconium is comparable to vacuum-arc melted material. In the case of selected titanium, zirconium, and uranium alloys, segregation is minimal in ESR material.

   (4) No significant metal refinement has been demonstrated yet as a result of electroslag remelting of reactive or refractory metals. An exception is oxygen reduction in molybdenum, chromium, and uranium.

(5) Electroslag melted titanium, zirconium, molybdenum, and copper exhibit comparable tensile properties to those of more conventionally melted material, although the impact strengths of electroslag melted titanium and zirconium are significantly lower.

(6) Electroslag remelting of aluminum shows no advantages over lower cost conventional melting processes unless significant inclusion removal, resulting in superior mechanical properties, can be demonstrated.

(7) Although ESR costs are higher than those for one vacuum-arc melting step (primarily due to greater energy consumption and slag costs), this may be nearly offset since single-melted electroslag ingots possess comparable quality to double or triple vacuum-arc melted ingots.

b. Recommendations

(1) Study the feasibility of preparing large ESR ingots of copper alloys, such as monel, chrome-copper, copper-bronze, copper-beryllium, and dispersion-strengthened copper using different types of electrodes, especially electrodes from powder compacts.

(2) Evaluate the various aspects of remelting chromium and its alloys by the ESR process, including characterization of the remelted material. Electroslag remelting may cause a lowering of the ductile-to-brittle transition temperature of chromium and certain alloys containing high contents of chromium.

(3) Examine the potential for electroslag remelting chromium to obtain chromium-based rare-earth alloys. With low (< 0.01 percent) nitrogen, carbon, and oxygen contents and optimum ($\sim$ 0.4 to 0.7 percent) additions of rare earths, grain refinement, improved ductility, and fine dispersoids have been observed (Rakitskii et al., 1973), although corrosion resistance was decreased (Tavadze et al., 1972). Such observations warrant further evaluations of electroslag melted rare-earth-chromium alloys.

(4) Study the feasibility for electroslag remelting uranium on a commercial basis because uranium appears to possess the best potential among reactive metals. This possibility exists because uranium has a lower reactivity and melting point than titanium or zirconium. Structural control using nonconsumable electrodes, shape casting, container vessel fabrication, and the AC electroslag production of U_3Si ingots for use in fuel pieces are areas which may prove fruitful. Explore further electroslag cladding with titanium for commercial application.

E. Electroslag Methods for the Manufacture of Large Ingots

1. General Comments

The present state of conventional ESR technology readily permits the manufacture of round ingots of up to 40 tons and slab sections of up to 25 tons. This assessment concerns ingot weights significantly larger than these sizes.

Presently available manufacturing routes may be divided into five general techniques:
1) Conventional ESR
2) ESR using strip/powder electrodes
3) Electroslag casting
4) Heavy-section electroslag welding
5) Central zone remelting process

Many technological problems are common to most of these methods.

2. Conventional Electroslag Remelting

a. Present Status and Anticipated Problems

The 110-ton capacity furnace constructed by the Leybold-Heraeus Company at Roechling-Burbach, Voelklingen, Germany, is the largest ESR furnace known to be operating (Choudhury et al., 1971). The next-largest furnaces are a 60-ton Soviet unit and the 50-ton Kobe Steel furnace. These furnaces produce ingots of cylindrical shape. Lukens Steel has the largest slab-section furnace (30 by 80 inches section, 30 tons capacity), although a 50-ton furnace (40 by 120 inches section) reportedly was ordered by Nippon Steel. Several ESR furnaces are capable of making ingots 5 feet in diameter, including the newest installation of Bethlehem Steel Corporation.

The principal technical problems that may be anticipated in large-scale operations are:
1) Permissible melt-rate and melting efficiency
2) Power supply
3) Ingot thermal regime
4) Slag selection for chemical requirements
5) Electrode manufacture.

b. Permissible Melting Rate

Since the melting rate directly represents furnace utilization and, hence, the installation's amortization rate, the highest possible melting rate compatible with acceptable ingot structure must be obtained. A generally quoted guide for melting rates in smaller ingot sizes is:

melt rate in 2.2 lb/hr (kg/hr) = ingot diameter in 0.04 inches (mm),

when the pool depth approximately equals the ingot radius. Such a pool depth normally leads to an acceptable ingot structure. Some rationalization exists for such a relationship, based on a constant heat transfer condition for any given ingot point as the ingot size increases. Such a value does not, however, provide an adequate basis for the design of a large, expensive furnace. The alternative solution is to use a heat-transfer model of the ingot to predict the entire ingot thermal regime as a function of time, size, and melt rate. Various attempts have been described (Mitchell et al., 1973) and an example of the predicted ingot pool shape is shown in Figure 12. At present, only very qualitative guidelines concerning what constitutes acceptable liquid-solid distance in terms of solid ingot properties are available. However, in very large ingot diameters, the maximum permissible melt rate cannot be defined by one simple relationship to ingot diameter. The normal melting sequence involves a starting, melting, and hot-topping period. In a small diameter ingot, the normal melting period is long with respect to the other two. In a large diameter ingot, this is no longer true since the commonly used dimensions require an ingot that is relatively shorter with respect to its diameter than the proportions of smaller ingots. Figure 13 shows this effect schematically. Obviously, there is no single melt rate for any one ingot. While it is legitimate to quote melt rates for small ingots, the maximum permissible melt rate in large ingots is a sensitive function of ingot length and must be followed closely if maximum productivity is to be attained. At present, few, if any, melters are able to do this on existing equipment.

The melting efficiency is the amount of power required to process a unit of metal. A commonly achieved value of this is 0.6 kWhr/lb (1.32 kWhr/kg), although values of up to twice this figure are quoted for low-resistance slags. Probably, 0.6 kWhr/lb will be observed in large ingots

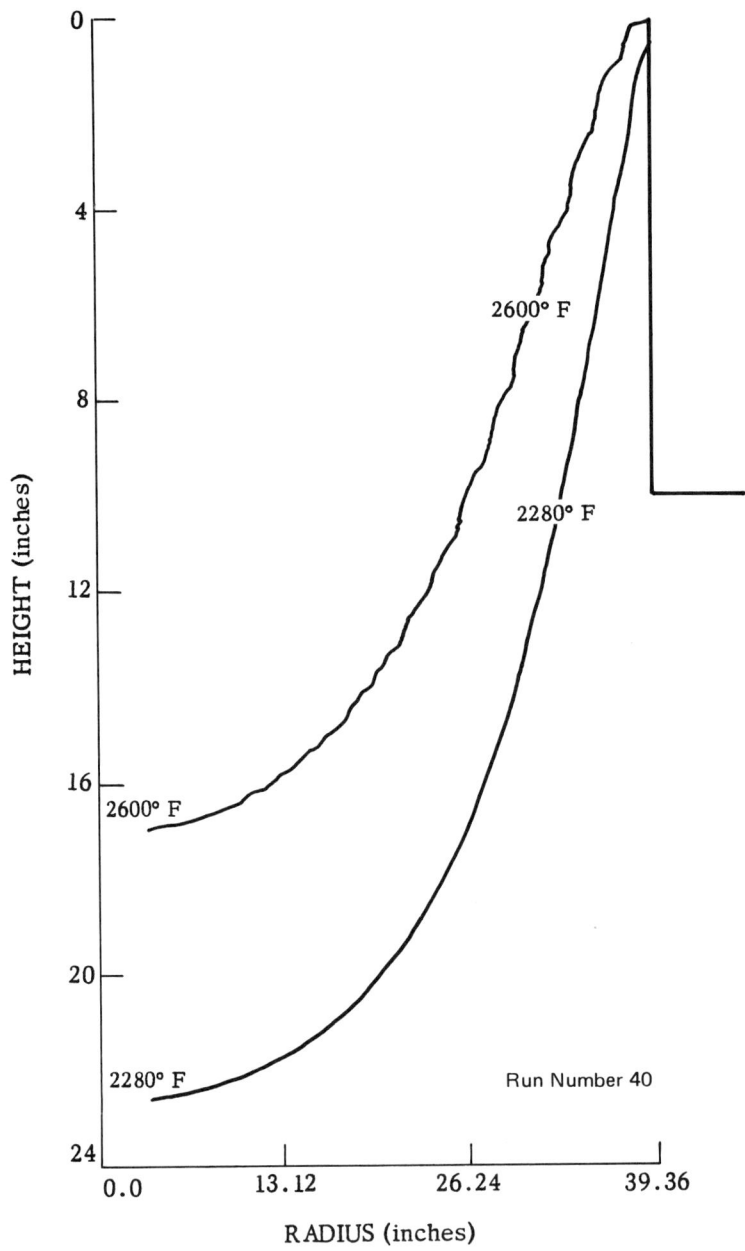

FIGURE 12 Thermal Profile for an 80-Inches-Diameter Ingot Melted in a Collar Mold at 4,400 lb/hr (from Mitchell et al., 1973).

Liquidus temperature (T_L) = 2610° F; solidification temperature (T_S) = 2280° F.

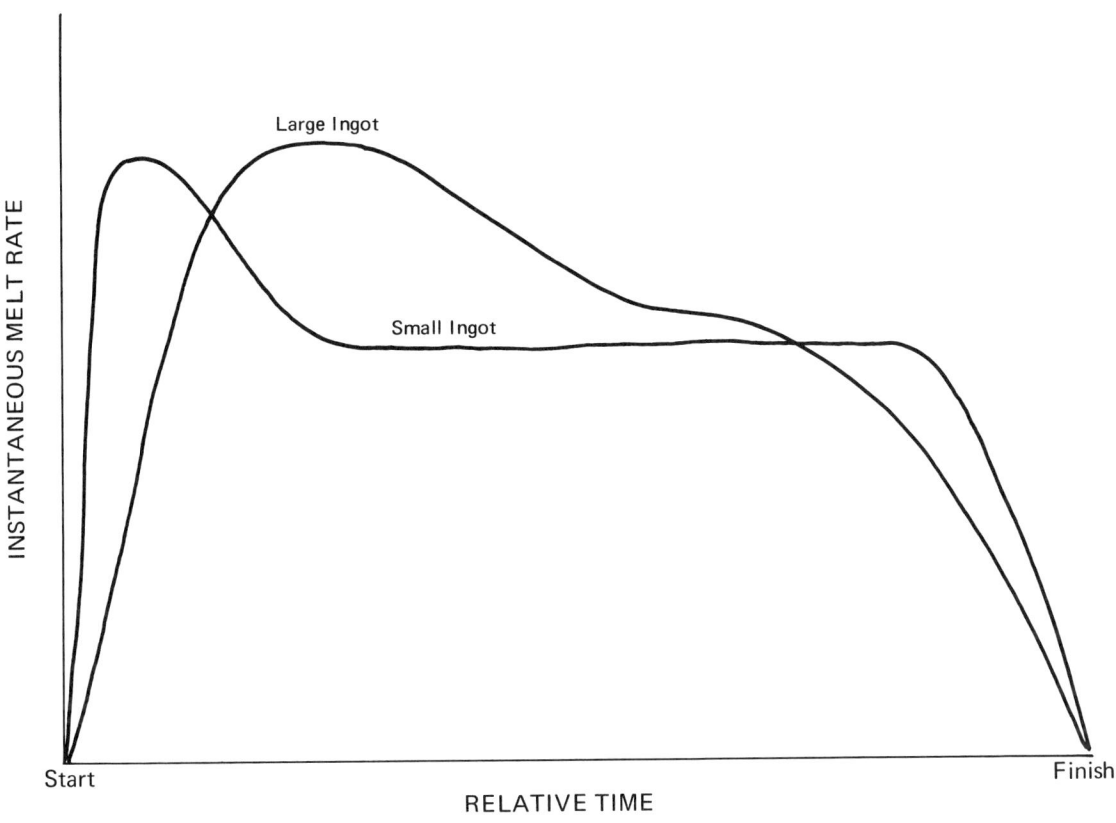

FIGURE 13 Instantaneous Melting Rate as a Function of Time for Two Diameters of ESR Ingots (from Mitchell et al., 1973).

NOTE: Melt programming is much more important in the case of a large diameter/length ratio.

although the scale-up calculations necessary to confirm this value have not yet been done. The importance of Figure 13 in defining the furnace requirements is now evident since the combination of the melting efficiency figure and the maximum permissible melt rate define the furnace power requirement and, hence, the installed power supply. For example, a maximum melt rate of 11,000 lb/hr (required initially in a 10-feet-diameter ingot to maintain an average of 6600 lb/hr from start to finish) requires a 6 megawatts, or approximately 7 megavolt amperes capacity. A maximum melt-rate of 6600 lb/hr requires 3.6 megawatts, or 4 megavolt amperes capacity. Hence, a furnace must be designed around the maximum permissible melt rate and not the steady-state average. The essential need for a careful study of the process heat balance in large ingots lies in this requirement.

c. Power Requirements

There are several possible power modes for use with large furnaces, most of which have been tried in practice (shown schematically in Figure 14). As mentioned above, the furnace power requirements are likely to be large and, since the inherent impedance of the ESR process is low, a low-voltage (40 to 60 volts) and high-current (> 100,000 amperes) method must be used. Most of the designs shown in Figure 14 attempt to ease the problem of supplying this power. DC melting has been established as the cause of chemical changes during processing, leading to composition changes between the electrode and ingot that are often deleterious to ingot properties. For this reason, most ESR furnace installations use alternating current of a frequency greater than approximately 4 Hz (cycles per second). It is reasonably certain that frequencies greater than 4 Hz are sufficiently reversible in the electrochemical sense to cause no frequency-dependent chemical changes. Consequently, there is no apparent reason, other than one of circuit design, for choosing a particular supply frequency.

The remaining major difference in the modes lies in the choice of number of phases and number of electrodes. Single-phase single-electrode methods are commonly used as is the bifilar configuration. Regarding large installations, the chief unknown factor in the single-phase single-electrode method is the division of current flow between the crucible return and ingot return paths. It has been reported that this division affects melting characteristics (e.g., surface quality and melt-rate) that may be very important in large ingots, but the precise mechanism is not known. The bifilar method carries an additional uncertainty in that the ingot must be connected electrically to the transformer neutral point. This connection represents the out-of-balance in melting current between the electrodes. The precise effect of variations in this component is not understood, but its control appears necessary to achieve melting stability. The operation of a bifilar installation has been described (Boucher, 1972).

The three-phase three-electrode version of the ESR process is in operation in several installations, and its main advantage is the balanced three-phase load to the power supply. The power supply is of traditional and well tested design and the technology of high-power installations is well established. The electrodes are driven independently, which removes the necessity of electrically defining the center point of the delta circuit.

In large installations, the efficiency of the power supply to the melting region is crucially important. Data available in the literature indicate that there is little apparent difference in electrical efficiency between the multi-electrode designs and a good single electrode design. The efficiency of bifilar installations is not higher than either of the above designs.

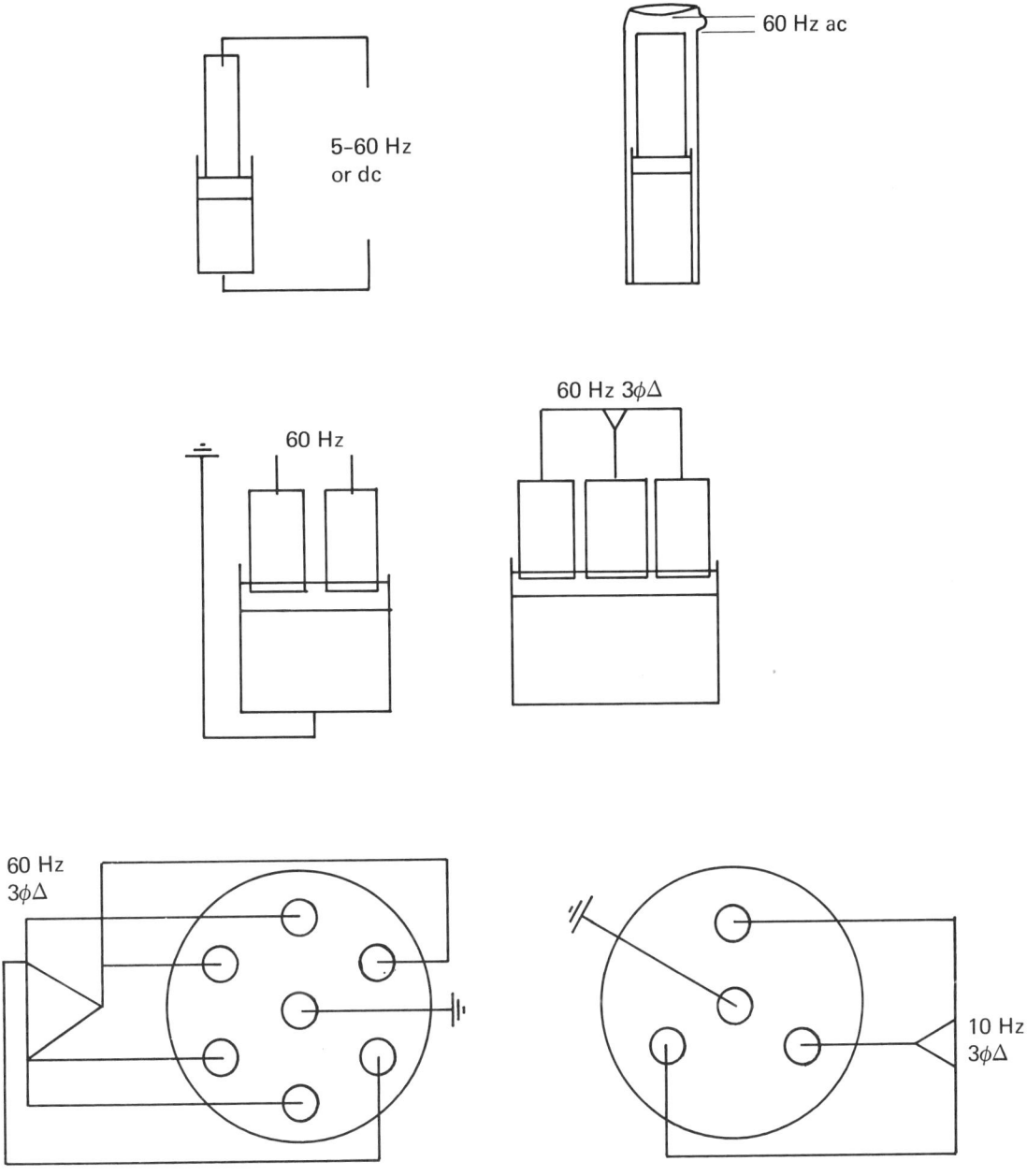

FIGURE 14 Electrode Geometry Configurations and Power Input That Have Been Used for Melting Large ESR Ingots.

d. Ingot Thermal Regime

The form of the solidifying region and the permissible melting rate have been discussed above, and the entire thermal regime of the ingot now must be considered as a function of the variables outlined.

The first major constraint to be considered is the minimum permissible ingot temperature. When large heat-treatable ingots are made conventionally, the ingot temperature is not permitted to fall below 750° F during its first cooling (i.e., in the as-cast state), and ESR ingots apparently also must be made following this practice. Since the ingot base is not only the first part to solidify but also strongly cooled, the question is whether or not the ingot base will attain too low a temperature. Using the models developed and described above, the primary factors influencing the base temperature are:
1) Ingot length
2) Melting rate
3) Base cooling mode
4) Hot-topping and stripping time

Ingot length obviously has a direct bearing on the base temperature since the heat supply is being removed to a point progressively more distant from the base. The melting rate is taken in conjunction with the base temperature, because the process is essentially unsteady-state and the base temperature is a function of melting time as well as ingot length. The base cooling mode controls the axial heat flow in the ingot base. The process termination steps are a most important factor in the overall process time and influence the base temperature through this variable. A computer calculation of the temperature regime at the base of an ESR ingot of 40 inches diameter by 160 inches long indicates that the base temperature falls to an unacceptably low value in a 40-inch diameter ingot when:
1) The base cooling exceeds 0.001 cal. $cm^{-2} s^{-1} °C^{-1}$
2) The melt rate average is less than 2200 lbs/hr (1000 Kg/hr)
3) The ingot length is greater than 160 inches (4 m)

Similar exercises on larger ingots have indicated a great need for care in this area. Since very little or nothing is known of the true average base heat-transfer conditions, that subject should be well understood before large ingots are attempted.

An equally critical area in large ingots is the choice between a moving or a static mold. Moving in this sense implies relative motion between the ingot and the mold. To operate a moving mold successfully, a strong ingot shell at the point of the ingot exit below the mold is required to contain the liquid. This need for ingot shell strength at the mold-ingot exit is an upper limit to the melting rate sustainable in a moving mold. This upper limit may be sufficiently high, but if it proves to be too low to satisfy either of the conditions of economics or base temperature, the design will not be feasible. In the context of both ingot shell and base temperature, the possible effects of ingot forced cooling or insulation must be considered. Air-cooling of the ingot has been reported as a means of influencing solidification structure in small (8 inches ⌀) ingots. At high melt rates, it could strengthen the ingot shell below, the mold; however, reheating the shell below the cooled region also might introduce thermal stresses that could fracture the ingot. To alter the bulk solidification, consideration may be given to cooling the upper part of the ingot and insulating the lower region to maintain the base temperature. These procedures will be substantially less effective in large ingots than in sections less than ∼ 12 inches because of the transfer of major heat-flow resistance progressively into the ingot interior. However, both procedures should be evaluated carefully.

The remaining question in thermal regime is whether or not an electrode change should be permitted. The controversy on this subject is extensive but the documentation is sparse. Ingot defects directly traceable to electrode change procedure have been found in several grades of steel and nickel-base alloys; however, the method is used routinely in many grades without apparent rejections. The controversy is restricted to small ingot sections (up to ∼ 24 inches). An examination of the change in thermal regime during a change of electrode in an 80-inches-diameter ingot does not indicate the likelihood of a detectable solidification defect after a normal change occupying 30 seconds or so. In a multi-electrode process where almost normal melting is maintained during a change, the effect will be even less noticeable. From a solidification standpoint, there is no reason to doubt the utility of electrode change methods in large ingots.

e. Chemical Requirements

Because of the very high heat-flow rates, it is exceedingly uneconomic to run an ESR furnace at less than maximum melt rate for chemical reasons. The desirable situation is to attain the maximum melt rate with the best possible refining.

Although the chemical problems faced in large ingot manufacture are not the same as those in small high-alloy ingot production, they are not less complicated. The most important requirement to be met by the process is chemical homogeneity. The results of analysis surveys on large ingots show that very little bulk segregation exists even in the hot-top region. The effective segregation coefficients of all elements lie in the range:

$$0.8 < k_{eff} < 1$$

The process acts as an efficient mixing reactor, not as a single-stage zone refiner. Also, the microsegregation was small, following the primary dendrite spacing as a simple function of local solidification rate. All examined ingots were columnar and directional in structure and, with few exceptions, free from macrodefects. On the basis of the published evidence, the macrostructural characteristics of the large ESR ingot should be much superior to the air-melt and static-cast ingot in every respect; however, it is not yet clear whether such a large improvement in structure is really an industrial necessity.

Several workers have shown that macrodefects of the "A" segregate variety can be produced in large ESR ingots under adverse conditions. These defects were thought to arise in essentially the same way as in conventional ingots — i.e., they result primarily from too wide a solidification zone and gas evolution during solidification.

The role of the slag in influencing solidification quality is essentially indirect — i.e., providing stable melting conditions at an appropriate melt rate. Thus, provision of stable melting has been defined as one of the boundary conditions for usable slag composition. This requirement generally appears to be of the same nature in large ingot manufacture as in small; however, the following aspects become more critical:

1) Suitable resistivity, viscosity, and interfacial tension
2) Suitable phase relationships

The precise numerical values of the former parameters depend on the melting method used. In the case of a small electrode/mold diameter ratio (fill-ratio), high-resistance low-viscosity slags must be used to generate adequate uniform heat (the slag composition "1/3 CaF_2, 1/3 Al_2O_3, 1/3 CaO" is an example of this class. In the case of a large fill-ratio, lower resistivities are permissible if adequate current capability is available. In configurations of extended path length (bifilar or three-phase), lower resistance slags may be used without requiring excessive current. Several good reviews are available on the topic of slag physical properties (Winterhager et al., 1970). Small relative changes in slag volume, resistivity, process temperature gradients, etc., are details that do not affect greatly either the operation or the economics of the small-scale process, but they have a profound bearing on the large-scale furnace. A small change in operating slag resistance, for example, may result in a 10 percent loss in melt rate. While this effect may be insignificant on a small scale, it may extend the process time on a large ingot to the point where base cracks become important. For this reason, methods must be developed to model the slag region and to monitor it during processing.

The slag phase composition is important for two main reasons: (1) melt-down time, and (2) slag skin composition. The slag melting process is essentially one of solution according to the phase diagram. Hence, a $CaF_2 + Al_2O_3$ mixture liquifies very slowly while a $CaF_2 + CaO$ mixture liquifies rapidly. The slag skin controls the mechanical stability of the melting region and, thus, must have adequate high-temperature strength as well as the requisite thermal conductivity. Little is known of the details of this process area. The slag also must distribute heat uniformly over a large electrode cross section. Electrode sections up to 40 inches are known to melt uniformly; however, no model or experimental evidence is available to describe the effect of local thermal disturbances on the electrode melting in very large sections.

The main chemical refining reaction to be utilized in large ingot practice is sulfur removal. The optimum sulfur level in the ingot depends critically on the end use. Evidently, a plate use would require lower sulfur levels than a large forging. An 80 percent desulfurization at electrode levels of 0.01 percent by weight sulfur is presently feasible without sacrificing too much resistivity by adding CaO to the slag composition. Such an ingot sulfur level is probably adequate for most purposes envisaged. In spite of the large metal/slag ratio, published calculations indicate that this level of desulfurization should be retained in large ingot furnaces.

In achieving this level of desulfurization, however, slag basicity must be at least 5 (implying a CaO activity approaching 0.5). In small ingots slag basicity represents no problem with respect to hydrogen transfer but this is not the case in large ingots. A consideration of the known characteristics of hydrogen in slags indicates a very high level of water activity at a basicity of 5. It is not surprising, therefore, that many reports in the literature describe unacceptable hydrogen levels in large ESR ingots. In spite of claims of electrolytic dehydrogenation treatments and slag treatment during processing, there is no clear evidence that hydrogen can be made to leave an ESR slag by any route other than solution in the ingot. The hydrogen content of the ingot is related closely to electrode hydrogen content, initial water content of the slag, and atmospheric humidity. The constants published for such relationships clearly show that melting a vacuum-degassed electrode through a slag (where $a_{CaO} = 0.5$) exposed to ambient air of 68° F dew point leads to unacceptable ingot hydrogen levels.

The solutions to the dilemma are varied. One obvious amelioration is to remove the lime component of the slag and accept little or no desulfurization. Another is to melt under a dry atmosphere. Clearly, more must be known about the kinetics and thermochemistry of the slag/water reaction before rational decisions can be made. It is equally clear that the control of hydrogen is a major and universal problem in large-ingot ESR practice.

The control of minor alloy elements (e.g., vanadium, aluminum, silicon, oxygen, nitrogen) also is noted in the literature as a point of difficulty. The level of oxidizable elements (principally silicon and manganese) may be controlled by continuous slag deoxidation, but this necessarily means an increase in the ingot aluminum content when aluminum deoxidation is used. This aluminum content appears primarily as the oxide or nitride, with the nitride causing difficulties in subsequent treatment steps in many steel grades. Very few deoxidation methods, other than aluminum, have been tried although the Soviet literature indicates that both calcium and rare-earth silicides may be substituted.

Aluminum control also is a problem when $CaF_2 + Al_2O_3$ binary slags are used to retain low hydrogen levels. In such a case, the Al_2O_3 activity is very high and a maximum aluminum pick-up is experienced. Japanese work on forging grades indicates the possibility of an additional 0.05 percent by weight aluminum arising from this source.

The control of oxygen is not a problem in large ingots. The inclusion distribution is invariably very much better than in equivalent conventional material and, apparently, there is a complete absence of inclusions greater than 0.002 inch (50 μm) in dimension. It is not likely that such inclusions would affect the necessary level of large ingot properties.

f. Electrode Manufacture

As noted above, there are certain restrictions on the effectiveness of ESR in large ingots. Essentially, an optimization is sought between hydrogen control, melt rate, and chemical refining. If any of these steps can be transferred to the electrode manufacturing stage, this aspect should be examined. For example, consider the cases of hydrogen and sulfur. The normal level of desulfurization attained in ESR varies from 30 to 50 percent in $CaF_2 + Al_2O_3$ slags to over 90 percent in $CaF_2 + CaO$ slags. However, $CaF_2 + CaO$ slags cannot be used either in humid atmospheres or in cases where a high melt rate is desired. In these latter cases, a $CaF_2 + Al_2O_3$ slag must be employed, and the electrode material produced at an adequately low sulfur level. In the case of hydrogen, if a very low hydrogen content in the ingot is required, the electrode material must be degassed. The economic viability of ESR processing sequences is not well defined. Specific objectives to be achieved in ESR should be investigated, especially in the light of new steelmaking processes incorporating the refining of liquid steel in the ladle.

Electrode manufacture is a particularly crucial step in determining the design and operation of the ESR furnace. The methods range from one electrode-one ingot to electrode change-multiphase procedures. Undoubtedly, if the melting capacity is available, the former method is the simplest with the highest fill-ratio and has the simplest configuration for electrode casting. But one of the more important economic advantages in ESR is that, in principle, large ingots may be manufactured without large melting and casting facilities. A number of publications discuss this point and suggest possible ways of synthesizing one large electrode from smaller segments. The principal difficulty is casting long thin segments and ensuring the mechanical and the electrical integrity of their assembly into a large electrode. If they are alternated head-to-base, no axial or radial segregation in the ingot is believed to result from this source.

The alternative method is to change electrodes and/or use a multi-electrode furnace. The main objection to the electrode change procedure appears to be twofold – i.e., the practice is subject to possible mechanical failure and the maintenance of a dry furnace atmosphere would be complicated.

Both objections are valid in principle but, as yet, there is no definitive evidence on the problem. Relatively small-scale experience (ingots up to 40 inches in diameter) indicates that there apparently is complete mixing in the ingot pool and, therefore, no radial segregation is expected to arise from small composition differences in electrodes being melted simultaneously.

In an effort to retain the advantages of multi-electrode preparation without electrode changing and without the necessity to cast long electrodes, the "tulip-mold" furnace has been proposed for large ingots (a section of a typical design is shown in Figure 15), but its operation is very difficult because

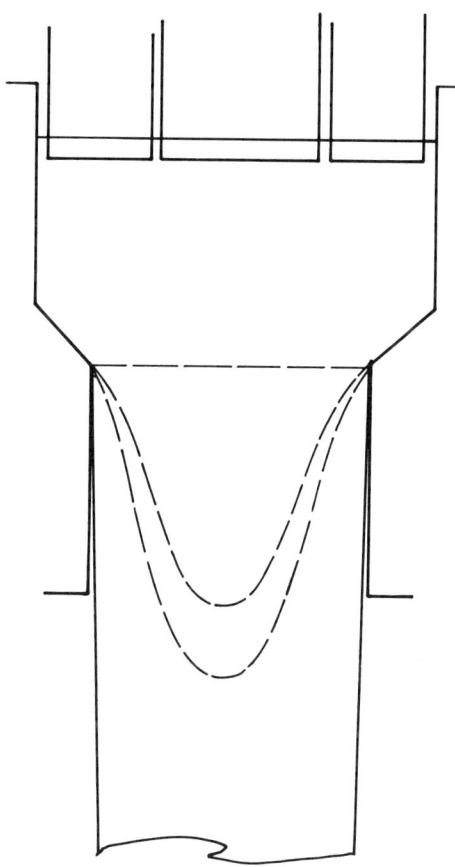

FIGURE 15 Multi-Electrode Reduced Section of Mold Configuration.

of the problem of accurately controlling the position of the slag/metal interface relative to the mold. The viability of the design in large ingot sections is not known presently and the largest such furnace is thought to be of 10-ton capacity although projections of up to 60 tons have been revealed.

3. ESR Using Strip/Powder Electrodes

As first proposed, the method of electroslag remelting using powder/strip combinations as feed material was known as the Kellog process. Regarding the large ingot installations, only the Cockerill-Ougree in Belgium is known to be in operation. The principle of the electroslag powder melting process is described in section VII.A.3 (Figure 32). The furnace was constructed by the Arcos Corporation and has been operated routinely for several years making forging-grade steels in cylindrical sections up to 40 inches in diameter and 160 inches in length (Descampes and Etienne, 1973). Cockerill-Ougree subsequently purchased two furnaces of conventional electrode-change design from the Boehler Company.

Basically, the furnace can be used as a three-electrode three-phase or a single-electrode single-phase furnace. For large sized ingots, the triple configuration is proposed (see Figure 34). The strip (approximately 4 by 0.16 inches [100 by 4 mm]) is fed by current-carrying rollers through a powder-feed device and into the slag. As in the equivalent welding equipment, the "stick-out" of strip between the current contact and the slag is an important operating variable.

The main problem areas in large ingot operation focus on:
1) Strip/powder quality
2) Feed continuity
3) Oxidation
4) Mold design

a. Strip/Powder Quality and Feed Continuity

The major advantage of this process is that it functions without large solid electrodes; however, it does require large quantities of strip and powder of a suitable quality. Since alloy adjustment is made through the powder component, the strip is usually a plain carbon steel with a suitably low sulfur content. The powder component presents the major difficulty. Most iron-base metal powders are available only in large quantities in high-oxygen grades. If high-oxygen powders are used in the Arcos process, heavy oxidation results with consequently high ingot oxygen contents. Powders with low oxygen contents are expensive and difficult to handle in bulk without reoxidation. With a large level of slag deoxidation and low oxygen feed, the process is capable of making satisfactory ingots; however, the penalties of heavy deoxidation are that the process is potentially difficult to control and that the maintenance of consistent slag conditions is virtually impossible.

The presence of higher-than-normal ingot oxygen contents leads to potential problems in meeting mechanical property specifications: (1) through the possibility of microporosity being created during solidification, (2) through the presence of a larger number of inclusions, and (3) through the effect on sulfides. These sulfides are found to be Type I in high-oxygen forging ingots, as opposed to Type III in well-deoxidized ingots. Such inclusions have a deleterious effect on both fracture toughness and fatigue resistance.

The maintenance of feed homogeneity is also difficult due to the necessity for retaining a close control on all feed components continuously. The strip acts as a magnetic carrier for the powder, and during the operation, large lumps of powder have been observed to fall from the strip and to be entrained in the ingot as composition inhomogeneities. The strip must be fed into the process with a high degree of stability otherwise intermittent arcing may detach small pieces of strip that also may appear as ingot inhomogeneities.

b. Oxidation

The problems of high oxygen feed have been outlined above. While this aspect may be avoided largely by using suitable powder compositions, oxidation during processing is significantly more difficult to handle. The oxygen source is the atmosphere and its route into the process is through powder oxidation during melting. The powder oxidizes as it is heated on the electrode and as portions of it circulate within the slag during melting. As with the feed-powder oxygen content, the oxygen from this source may be reduced by adding sufficient deoxidant but at the expense of a change in slag composition and bulk.

c. Mold Design

The mold design in any cold-crucible process presents several difficult problems. The cooling action is essentially by a combination of boiling and nonboiling water. The available studies on heat flow in ESR molds indicate that a significant part of the high-temperature mold region is in the boiling regime (Figure 16). At the implied high heat fluxes, a poorly designed water-flow system may convert easily from nucleate boiling to film boiling as large areas of the mold become covered with a steam film. The purpose of high flow rate cooling designs is to reduce the possibility of this happening, and, hence, to maintain a low overall mold temperature. The water flow rate has very little influence on the overall heat-flow resistance until film boiling takes place. The mold/water interface usually does not rise above 300° F, but the mold inner surface may be above 570° F for a significant time during the melting process. In a static mold, this high-temperature region moves progressively up the mold, but in a moving mold it remains essentially in the same place throughout the melt.

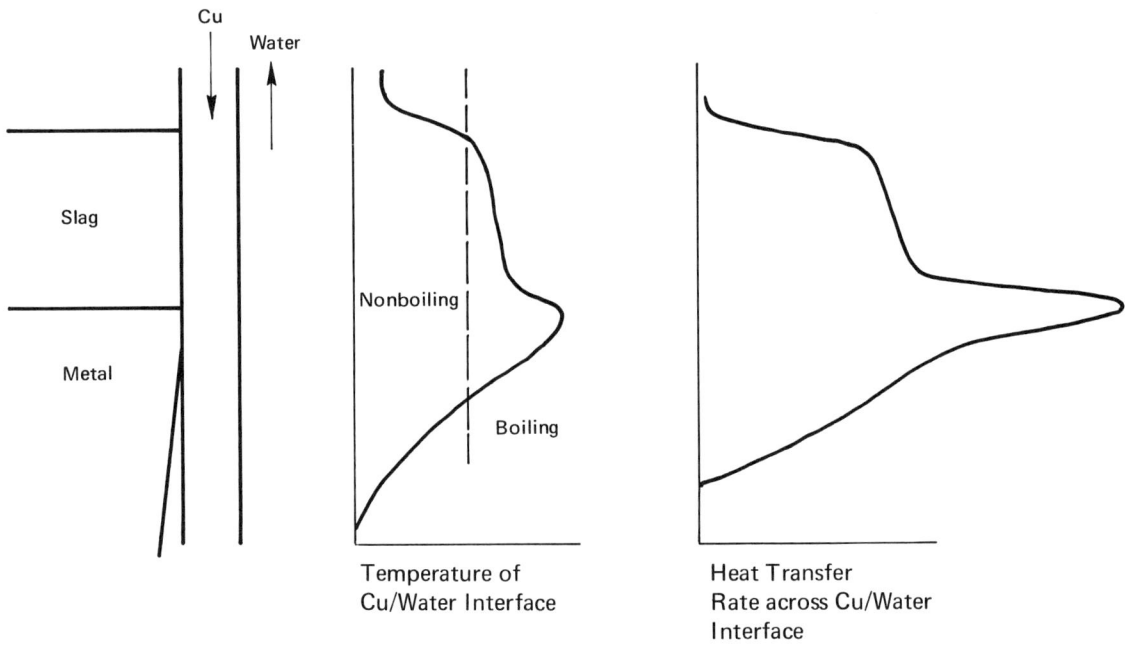

FIGURE 16 Heat Transfer Regime at the Copper/Water Interface in the Process Region (from Mitchell et al., 1973).

The basic problem of ESR mold design is to allow for this temperature regime. Copper and most copper alloys creep extensively during such thermal cycles, and the resulting mold deformation represents a most serious challenge to the designer of large molds. In the powder/strip furnace, the problem is compounded by the presence of lumps of powder circulating in the slag. These lumps intermittently cause arcing on the mold wall and local erosion and distortion. In extreme cases, this phenomenon also leads to local puncture of the mold. Various solutions have been proposed, including splitting the mold horizontally with an insulating layer, but the problem essentially remains unsolved.

d. Power Supply

In the large ingot configuration, the powder/strip furnace would require a three-phase power supply and, therefore, would offer the advantage of balance between the phases. Because of the very low effective thermal capacity of the electrodes, the power supply must accommodate rapid changes in demand, which may prove difficult in a large furnace.

e. Ingot Thermal Regime

There is no reason to suppose that the thermal regime in the powder/strip furnace will differ significantly from that of other furnaces operated at the same melt rate. Since a moving-mold or ingot-withdrawal configuration must be used, the problem of the ingot shell strength at the mold-exit point could be encountered above the connection.

f. Scale-up Problems

Apparently, the main problem that this method faces in scale-up is the electrode. There is clearly a maximum size of strip that can be handled conveniently and to scale up the process either more strip-feeding heads must be provided or the strip must operate at a higher feed rate. It is not presently known which of these alternatives is proposed, but the availability of powder of adequate quality on a sufficiently large scale must be considered.

A purely technical problem in production of a large ingot is the radiative heat loss from the slag surface. The effective fill-ratio is exceedingly small and, hence, most of the slag surface is free to radiate. The effect of this loss on melting efficiency has been examined on a small scale but it would represent a serious heat loss on a large scale. In addition, difficulties could be encountered in protecting equipment against such a large exposed radiant source and appropriate covers would be required.

4. Electroslag Casting

a. Present Status

The idea of using electrical power as a heat source for hot-topping has been investigated very widely. The specific area of electroslag hot-topping, using a consumable electrode, has been used in several meltshops. The technique of using this method, or a consumable arc electrode, with or without vacuum application, has been the subject of several patents. The technique of casting liquid metal through a slag in an uncooled mold also is well established. The logical combination of these methods, using a liquid metal feed and an electrically heated slag, is known as electroslag (ES) casting. There are several literature reports of studies and patents in this area that originated almost entirely in the USSR (Kamenskii et al., 1971; Belous 1958; Chernega, 1959).

The process is attractive since it eliminates the need for both a large solid electrode and a large power supply; however, in its application to large ingots, technical problems concerning pouring technology, powder supply, mold design, and chemical refining are envisioned. These areas are considered below.

    b. <u>Pouring Technology</u>

The presently reported technique of electroslag casting is based on supplying power to the slag while simultaneously slowly pouring in liquid metal. However, this is only one of several possible modes. The power may be supplied by consumable or nonconsumable electrodes and may be applied either during or after pouring. In the cases where the power is to be applied during pouring, the metal stream must be regulated carefully over a significant time. The reported case of a 7-ton ingot indicated 20 to 30 minutes casting time. Evidently, when the metal is added at this rate, a controllable valve must be used to supply the metal. The technology for this is probably available but has not been detailed in this application. At high rates of liquid metal addition, the internal heat-transfer mode in the solidifying ingot will be substantially different than that in an ESR ingot. If the ESR rate is desired, both metal and power must be added at a rate that simulates the ESR method. It is highly unlikely that a combination of external melting plus ES casting, at the typical rates of ESR, will represent a significant energy savings. The advantage in that mode therefore would lie primarily in the elimination of electrode casting.

On the other hand, in very large ingot sizes, the ESR metal quality probably is unnecessarily high. If this is so, ES casting at much higher rates than ESR represents a potentially useful compromise. The essential advantages here are in the elimination of electrode casting, a higher production rate, simpler machinery, and much lower power consumption. The technological question, as yet not resolved in the literature, concerns what pouring rate will result in an acceptable ingot structure.

    c. <u>Power Supply</u>

If the melting and superheating energy of ESR (approximately 550 kWhr/ton) are removed and if the casting rate is increased simultaneously, the required power for maintaining directional solidification is reduced greatly.

The values of supply, quoted for casting at the 25-ton-ingot size, are in the range of 500 to 800 kilovolt amperes. The requisite power supply for the casting machine is thus very much smaller than for the equivalent ESR furnace. The overall melting energy of the ingot casting process is reduced by at least a factor of 2 (from 1300 → 600 kWhr/ton) even if the energy cost of an ESR electrode is considered zero.

The construction of the electrical supply system is much simpler than in ESR; however, the problem of transmitting the power to the working slag must still be faced. Soviet literature is somewhat ambiguous on this point, but apparently both consumable and graphite electrodes have been used. In principle, with a sufficiently low CaO activity, graphite electrodes should be usable without significant carbon contamination. No evidence has been reported on this point since the ingot compositions used to date are mainly steels with at least 0.2 percent by weight carbon. There is no definitive evidence on the optimum mode of power connection since the literature quotes both monofilar and bifilar connection modes.

d. Mold Design

Mold design necessarily will depend greatly on pouring time. If the time is long enough to establish a large axial heat flow, the mold will have to be water-cooled. If the time is shorter, the heat may be accommodated partly by accumulation in a thick conventional mold and partly by convective dissipation from its outer surface. Published data indicate that this point has been difficult to decide. The early Soviet work was performed either with full water-cooled molds or with water-cooled segments; however, later studies have been performed with conventional molds, probably cast iron. Details of the base plate design have not been given. At present, while the link between the pouring rate and the mold design may be known, the data are not sufficient for decision making.

e. Chemical Refining

The chemical refining action of conventional ESR has been attributed largely to the breaking-up of the metal flow into droplets of high specific surface, but this is quite incorrect since the droplet residence time is so short as to render its overall contribution very small. The main refining sites are the electrode film and the ingot/slag interface. If one of these sites is removed entirely and if the residence time at the other site is reduced significantly, in principle, the refining capacity of the system is curtailed severely. However, this broad conclusion must be placed in the context of the process considered. In the case of sulfur, the conventional ESR process has been shown to come close to equilibrium at the electrode/slag site. Hence, essentially all the refining capability of the slag is achieved before the metal reaches the ingot/slag site. Since in the projected ES casting residence times this latter site will have a specific reactivity that is comparable to the conventional ESR electrode site, a refining reaction for sulfur is anticipated in ES casting that is close to that of an ESR furnace. Unfortunately, the same should be true of hydrogen; therefore, a major difficulty is the compromise between desulfurization and hydrogen transfer.

It has been shown that most of the ESR electrode inclusions are removed during the melting process at the electrode tip. Subsequent reactions in the solidifying ingot produce an entirely new distribution of oxide inclusions in the solid. Therefore, one must decide if the ES casting method offers the same potential in removing nonmetallic inclusions.

Reported work on the process of pouring metal into molds containing slag, but without additional heating, indicates that inclusions are not reduced sufficiently to justify using the process. By analogy, the same result for ES casting could be anticipated and indeed is documented in the literature on oxide inclusion contents. For small ingots of specialty steel, this is obviously an unsatisfactory situation; however, for large ingots, most of the sensitivity of mechanical properties to inclusions may arise from oxide inclusions in the size range above 0.002 inch (50 μm). In conventional ingots, such inclusions occur either by massive segregation on freezing or as exogenous inclusions. The former would not appear in ES casting because of the absence of such segregation, and the latter should be removed during pouring because of the high forces attracting them to a slag/metal interface. Thus, the literature seems to indicate that the degree of oxide inclusion removal in ES casting may be adequate for large ingot quality.

5. <u>Welding Methods</u>

   a. <u>Present Status and Problems</u>

The objective of electroslag welding method is to fabricate large ingots before forging by welding smaller pieces together. The two versions presently being used on a large scale are Soviet (Paton et al., 1973) and Czechoslovakian (Zeke and Zelko, 1969). Both reportedly have been tried in North America but are not practiced presently.

The essential difference between the two techniques is the way in which power and electrodes are introduced into the working region. The Czechoslovakian method is essentially a scale-up of the conventional consumable guide electroslag welding method; the Soviet version is a double bifilar slab furnace in which the two long faces of the slab are the ingot faces to be welded. The Czechoslovakian method has been used to make welds 120 inches thick by

80 inches high while the Soviet method has been used to join pieces 80 inches in diameter to a maximum weight of 160 tons. The reported problems in the methods involve:

1) Mechanical unreliability and complexity
2) Power instability
3) Contraction stresses
4) Composition inhomogeneity.

Another variant of this procedure is reported in the literature, in which the center of a large conventional ingot is removed by punching and the hole is filled in using a consumable electrode ESR process (Cooper, 1972; Paton et al., 1973; Heppenstall, 1973).

b. Mechanical Performance: Power Stability

A simple modeling exercise shows that since the heat flows in the welding region are extremely high, the process is very sensitive to changes in thermal regime. Generally, it is recognized that the small-scale ES welding method is very prone to form small regions of nonpenetration when welding conditions are unstable. The same phenomenon is observed even in large welds.

The main difference in operation between the two butt-welding techniques lies in the electrode configuration. In the Czechoslovakian method, the feed occurs as wire into a narrow gap (~ 1.6 inches) guided by insulated hollow plates that act as consumable electrodes. In a 10-feet-wide weld, there are 15 of these assemblies working simultaneously. Should one assembly fail or be temporarily interrupted, the loss of heating in that region is sufficient to cause an area of nonpenetration. In the Soviet method, the electrodes feed into a much wider gap (~ 80 inches) and consist of four plates, two fixed and two mobile. The power connection is double bifilar and the operating conditions are very similar to those in a conventional ESR slab furnace. This assembly has a much larger effective thermal capacity than the wire-feed assembly and the effect of changes in power supply or electrode feed consequently is reduced greatly. In addition, the assembly is mechanically more simple and less likely to fail.

It should be noted that like all other large ingot ES methods (except ES casting) these processes are an uninterruptible load and must operate continuously for possibly a 100-hour period.

c. Contraction Stresses

The progressive and unsteady-state nature of the welding processes causes contraction stresses to develop in the weldment. Clearly, if two 100-ton blocks are constrained while being butt-welded together, the weld very likely will crack. In the hole-filling method, the central cavity is filled with hot metal that contracts on cooling and, therefore, develops contraction stresses. The two production butt-welding procedures allow for this by offsetting one ingot with respect to the horizontal axis of the other. It is claimed that the hole-filling method can be operated without contraction cracks by suitably adjusting the melting rate. Evidently, this area is one in which investigations should be made to determine the magnitude of the problem.

d. Composition Inhomogeneity

The results reported to date on the mechanical quality of the butt-welding methods have been quoted from ESR base materials; therefore, one would not expect problems to arise from axial segregation, center porosity, or weld/metal mixing. In the welding of conventional ingots, the weld progresses in a direction that cuts across the axial segregation. Hence, difficulties could be met in maintaining weld quality across a region with porosity and extensive macrosegregation. There have been no reports on the degree of lateral mixing in the slab-shaped pool of such a weld and, at present, the degree to which mixing would homogenize the weld region cannot be judged. The mechanical evaluations reported on the ESR ingot welds indicate entirely satisfactory properties.

It is held that hydrogen problems do not appear in these methods because the volume of hydrogen-charged material is small compared to the volume of the final ingot. The hydrogen is thought to homogenize rapidly and produce a bulk content at an acceptably low value.

6. Central Zone Remelting Process

Midvale-Heppenstall Company has patented (Cooper, 1972) a consumable mold ESR process and has performed feasibility studies on small ingots and, more recently, on ingots exceeding 50 tons. Presently, Midvale-Heppenstall and Kloeckner-Werke AG at Osnabrueck, West Germany are developing the "MHKW process" jointly for the commercial production of large ingots. This process also is being studied by the Soviets (Paton et al., 1973).

First, a large conventional ingot is cast, upset forged, and its center is punched out. The hollow ingot then becomes an ESR mold into which an electrode of similar composition is remelted by conventional means, as with a single-phase ESR furnace. The inside diameter surface zone is melted back to a controlled degree and progressive solidification of the new "ingot center" takes place as depicted in Figure 17. In this fashion, the normally poor or marginal center portion of a large ingot is replaced with a sound, refined structure.

Although much work remains to be done, the process has shown sufficient promise that in 1974 Kloeckner-Werke AG installed a nominal 60-inch furnace with the capability for central zone remelting ingots up to 140 inches in diameter and 330 tons in weight.

If this process proves sound, it would have a profound impact on large ESR furnace design. Basically, ESR furnaces of conventional single phase line frequency and operating at modest power levels could melt new cores in ingots ranging up to 330 tons. Such a furnace would eliminate some of the formidable equipment and processing problems and capital costs that the equipment manufacturers face when attempting to build and operate a conventional 330 ton ESR furnace.

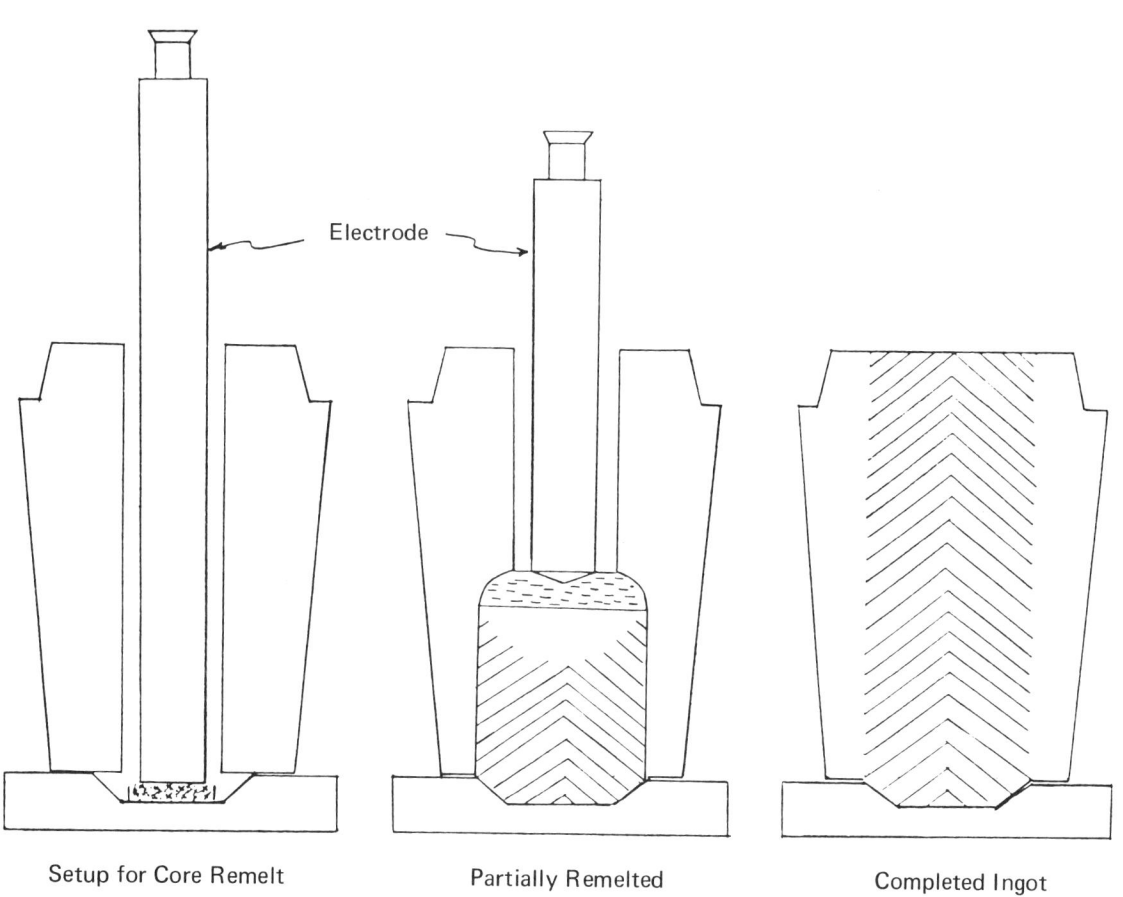

FIGURE 17 Central Zone Remelting Process (from Cooper, 1972).

Preliminary results on initial test ingots produced by the central zone remelting operation at Kloeckner-Werke AG were reported as follows (Cooper et al., 1974):

First ESR Melt No. 303254 was made on February 13, 1974.
    Original Melt No. 414362 – Type 26 Ni Cr Mo V 14.5 (F.R. Germany)
    Electrode Melt No. 114705 – Type 26 Ni Cr Mo V 14.5 (F.R. Germany)
    Ingot blank from top half of 137.5 short ton ingot upset forged and trepan punched for core remelting:
        Outside Diameter: 86.6 inches
        Inside Diameter: 29.5 inches
        Height: 77.2 inches
        Electrode Diameter: 15.75 inches
        Electrode Length: 271.6 inches

Chemical Analysis Results

Location	C	Mn	P	S	Si	Ni	Cr	Mo	V	Al
Ingot – Melt No. 414362 – Ladle	0.29	0.31	0.007	0.008	0.21	3.50	1.59	0.38	0.10	0.001
Center Core Punched from Ingot	0.30	0.29	0.007	0.006	0.20	3.55	1.62	0.36	0.09	< 0.001
Electrode – Melt No. 114705 – Ladle	0.32	0.30	0.009	0.013	0.30	3.76	1.59	0.37	0.08	0.001
Core Repunched after Core Remelt	0.30	0.28	0.008	0.005	0.06	3.69	1.70	0.36	0.08	0.001

After the completion of the Core Remelting Process, the ingot blank was again trepan punched on the press, in preparation for another core remelting operation. The core obtained by this trepan punching was analyzed with the results shown above. The drop in silicon content was not unexpected, since the slag was composed of CaO, CaF_2, and Al_2O_3. During the remelting operation, aluminum granules were added to the slag. The further core remelting operation on the same ingot blank was as follows:

Second ESR Melt No. 303256 was made March 5, 1974.
    Original Melt No. 414362 – Type 26 Ni Cr Mo V 14.5
    Electrode Melt No. 114728 – Type 26 Ni Cr Mo V 14.5
    Reforged and repunched ingot blank:
        Outside Diameter: 95.7 inches
        Inside Diameter: 29.5 inches
        Height: 72.4 inches
        Electrode Diameter: 15.75 inches
        Electrode Length: 341.7 inches

Chemical Analysis Results

Location	C	Mn	P	S	Si	Ni	Cr	Mo	V	Al
Ingot – Melt No. 414362 – Ladle	0.29	0.31	0.007	0.008	0.21	3.50	1.59	0.38	0.10	0.001
Electrode – Melt No. 114728 – Ladle	0.29	0.35	0.010	0.006	0.25	3.49	1.59	0.39	0.09	0.002

After core remelting, the ingot blank was forged to 63 inches diameter. A slice was cut from the top and bottom. Analyses positioned across the slices are as follows:

Chemical Analysis Results

Location	C	Mn	P	S	Si	Ni	Cr	Mo	V	Al
Top Slice										
#1 – Outer Zone	0.30	0.29	0.007	0.008	0.20	3.56	1.64	0.38	0.09	0.004
#2 – Junction of Remelted Zone	0.30	0.29	0.007	0.009	0.19	3.54	1.65	0.38	0.08	0.004
#3 – Remelted Zone	0.32	0.30	0.008	0.006	0.05	3.60	1.64	0.38	0.08	< 0.003
#4 – Remelted Zone	0.31	0.31	0.007	0.006	0.04	3.60	1.64	0.38	0.07	< 0.003
#5 – Junction of Remelted Zone	0.31	0.29	0.007	0.007	0.18	3.54	1.65	0.37	0.08	0.005
#6 – Outer Zone	0.30	0.29	0.007	0.009	0.20	3.54	1.60	0.37	0.09	0.003
Bottom Slice										
#1 – Outer Zone	0.30	0.29	0.007	0.007	0.18	3.54	1.64	0.35	0.09	0.005
#2 – Junction of Remelted Zone	0.26	0.22	0.007	0.007	0.07	3.62	1.58	0.35	0.07	0.001
#3 – Remelted Zone	0.33	0.14	0.007	0.008	0.01	3.62	1.53	0.36	0.07	< 0.003
#4 – Remelted Zone	0.23	0.16	0.006	0.007	0.02	3.50	1.50	0.34	0.06	< 0.003
#5 – Junction of Remelted Zone	0.29	0.20	0.007	0.009	0.01	3.57	1.58	0.36	0.07	0.005
#6 – Outer Zone	0.29	0.29	0.007	0.008	0.19	3.54	1.64	0.36	0.09	0.003

The remaining ingot blank was forged to a turbine disc. A core punched from the hub of the disc forging gave the following results:

Chemical Analysis Results

Location	C	Mn	P	S	Si	Ni	Cr	Mo	V	Al
Punched Core from Turbine Disc	0.29	0.24	0.008	0.007	0.03	3.56	1.56	0.34	0.08	0.003

The slag used in these trials was of the $CaO\text{-}CaF_2\text{-}Al_2O_3$ type. No efforts were made to maintain the silicon content in the remelted core at the same level as in the original ingot and electrode. This can be controlled readily to desired higher levels through slag selection and proper deoxidation control.

7. <u>Conclusions and Recommendations</u>

   a. <u>Conclusions</u>

   (1) ESR technology is sufficiently developed to permit the manufacture of 60-ton (60 inches ∅ by 120 inches long) ingots in round section and 30-ton (40 inches by 100 inches by 120 inches long) ingots in slab section on a production basis and 110-ton (92 inches ∅ by 120 inches long) ingots on a development basis. It is the only potential remelting method for high-quality ingot manufacture in sizes above 50 tons.

   (2) Two electroslag welding methods have successfully been used to fabricate large ingots – the Soviet method (bifilar welding) on 100 inches ∅ (160 tons) ingots and the Heppenstall method on 80-ton round section ingots.

(3) The major technical problems in developing the ESR process for larger ingots involve: hydrogen control; furnace and mold design and power supply; and process modeling for optimum melting rate, chemical refining, solidification behavior, and hot-topping cycle.

(4) No clearly recognized criteria concerning future requirements for large ingot materials are available to serve as the basis for a critical evaluation of the four available alternative electroslag processes – conventional ESR, ES casting, bifilar welding, and the Heppenstall process.

(5) The mechanical properties of large ESR ingots are clearly superior to those of ingots currently produced by the only alternative process – i.e., air melting and static casting.

b. Recommendations

(1) The kinetic and equilibrium factors governing hydrogen behavior in electroslag systems should be studied. Many of the basic data are available but lack application to the specific case of large low-alloy steel ingots. The chemical refining reactions should be investigated in an attempt to maximize slag heating efficiency, ingot surface quality, sulfur, hydrogen and oxygen removal, and alloy composition control. The problems of slag composition and control are common to all four processes.

(2) Adequate heat-flow models of all four processes listed in conclusion 4 should be developed. This will enable a real assessment to be made of: melt rates attainable, crucible design, hot-top cycles, process economics, power requirements, control system design, and solidification structure and segregation control.

(3) A systems approach should be taken to studying the feasibility of integrating one of the methods into large component manufacture. The component requirements should be defined by the user in respect to structure and composition at the forged and heat-treated stage. The precise requirements of the steelmaking sequence then can be defined and the relative merits of the following items assessed:

(a) Chemical refining in the ES process or in a prior process step

(b) A complete ESR ingot as opposed to one made by a composite method and also conventional method with respect to segregation and structure

(c) The economic role played by modifications permitted in the forging sequence on ESR ingots, as opposed to conventional ingots, or composite ingots; and by as-cast ESR structures

(4) Studies should be made to determine the maximum size required in ESR ingots, particularly in slab shapes. The study should evaluate the potential of component fabrication by joining several ESR ingots, possibly in the as-cast condition.

VI. NOVEL APPLICATIONS OF ELECTROSLAG TECHNOLOGY

A. Introduction

Recent advances in ESR technology provide a new method for greatly improving the quality of cast metal in comparison to that provided by conventional static casting.

The integrity of cast metal depends mainly on two factors: the purity of the liquid metal and the nature of the solidified structure. Purity of the liquid metal may be improved by processing under vacuum or by treatment of the metal with special slags or gas mixtures. Production of shrinkage-free casting can be accomplished through controlled solidification in cooled molds. The electroslag remelting and ingot casting process combines metal purification and ingot structure refinement functions in one operation. The flexible process features of electroslag refining and casting have been utilized to produce various shaped castings.

Novel applications of the ESR process include casting of square ingots, rectangular ingots using two consumable electrodes connected in series, fluted ingots, extrusion shapes, hollow ingots, near-net-shaped castings (such as rolls and spindles), and near-net-shaped complex castings (such as valve bodies, crankshafts and pressure vessels).

Other novel applications of the ESR process are the production of smaller section ingots using a larger section electrode in a funnel-type mold and the production of multiple ingots using one large electrode remelted in a funnel-type mold and multiple cavities in the bottom opening or the same number of multiple electrodes remelted in a moving mold arrangement so as to produce the same number of multiple ingots.

Incentives, objectives, and technical justifications for each of the various adaptations of the ESR process may be different. These considerations include factors such as the nonavailability of alternative material processing equipment, cost trade-offs among competing processes, improved quality and performance, and manufacturing schedule reduction.

The novel applications of the ESR process are mostly of a developmental nature and some are pilot-scale studies. Those that have reached commercial status include ESR casting of slabs, square ingots, roll bodies, and roll cladding. In the USSR and Japan, ESR casting of near net shaped components reportedly has reached commercial status; however, detailed reports of these process developments are maintained as proprietary information.

B. Novel ESR Process Adaptations

Novel ESR ingot casting concepts are:
1) Manufacture of small-cross-section ingots from large-cross-section electrodes using a funnel-shaped mold
2) Manufacture of multiple ingots from single or multiple electrodes using a funnel-shaped mold
3) Manufacture of rolls
4) Manufacture of multicontoured castings
5) Manufacture of slabs using the paired-electrode (bifilar) arrangement
6) Manufacture of hollow ingots
7) Manufacture of small and large pressure vessels with integral domes and side ports
8) Manufacture of large castings by the YOZO technique
9) Manufacture of ingots using metal composite electrodes
10) Manufacture of ingots or castings using molten metal feed in conjunction with electrodes
11) Manufacture of ingots or castings using metal chips (scrap) or metal powder feed in conjunction with either consumable or nonconsumable electrodes
12) Direct reduction of oxides and ores via ESR.

1. Manufacture of Small-Cross-Section Ingots from Large-Cross-Section and/or Multiple Electrodes

Manufacture of small-cross-section ingots from large electrodes is done in movable, funnel-shaped, water-cooled copper molds as shown schematically in Figure 18. The success of this technique depends on holding the molten metal level within the straight section of the mold so that a slag envelope will form on the ingot. It is this slag skin that provides a smooth finish on the ingot surface. The insulation provided by this solidified slag skin has a decisive effect on the metal solidification rate.

This adaptation of the ESR process involves coordination of several movements in the system; these movements include that of the electrode, the mold, the metal level, the slag level, and the ingot. The use of different probes to determine slag-metal interface level and temperature and a computer for melt-rate determination and metal transfer are highly desirable for industrial exploitation of this concept.

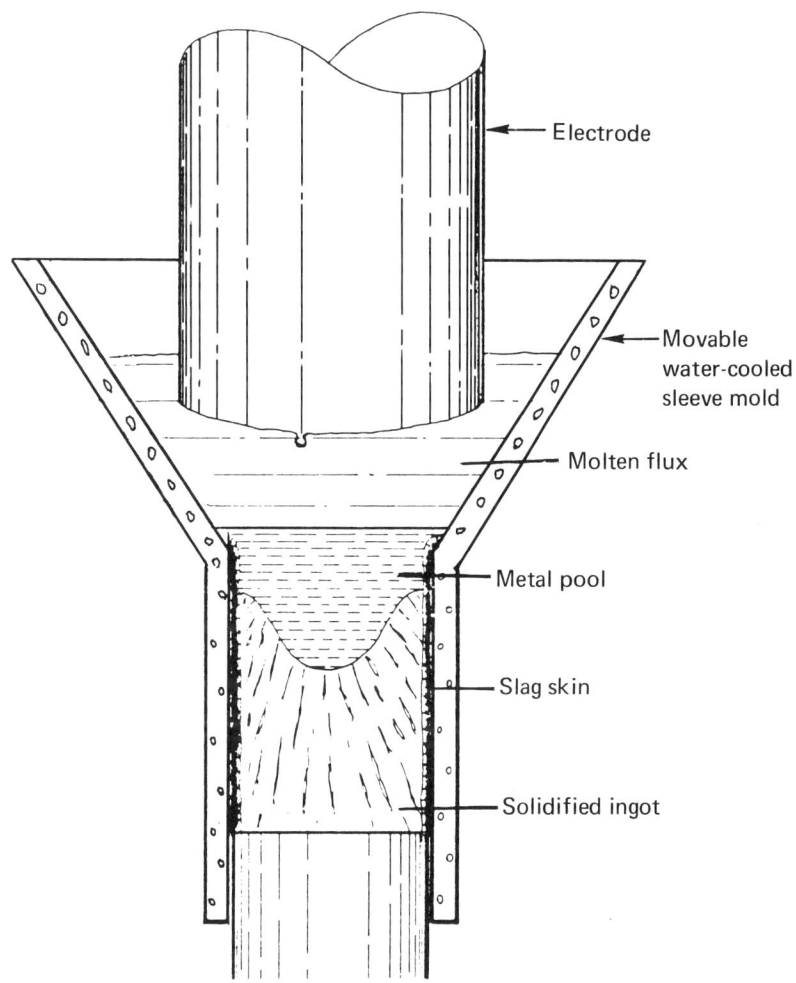

FIGURE 18 Manufacturing Concept of Small-Cross-Section ESR Ingots from Large-Cross-Section Electrode.

2. <u>Manufacture of Multiple Ingots from Single or Multiple Electrodes</u>

The logical extension of ESR melting in funnel-shaped molds is the manufacture of polyingots using one large electrode or a plurality of small electrodes. This arrangement is illustrated schematically in Figure 19.

When only one electrode is used, the electrical connections are simple. With multiple electrodes, the electrical power connections could become quite complex. This ESR arrangement of producing multiple ingots has industrial promise. The ingots produced could be used directly in many applications. The ingot could be round, square, rectangular, or any other simple shape.

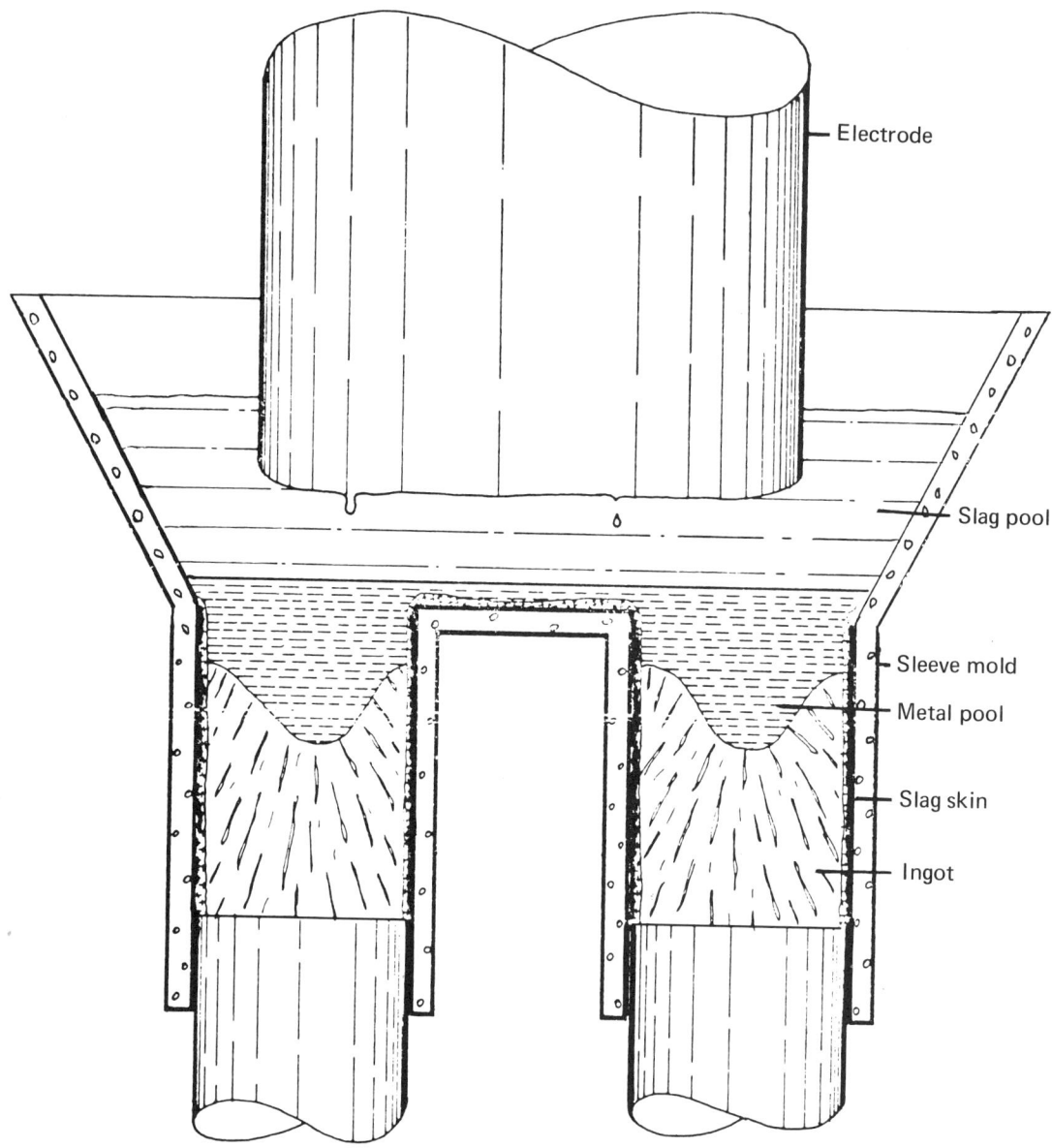

FIGURE 19 Manufacturing Concept of Multiple ESR Ingots from a Large-Cross-Section Electrode.

3. Manufacture of Rolls

The manufacture of roll bodies using ESR is a simple operation. Figure 20 schematically shows the steps involved. It is also possible to make composite rolls (Figure 21) or extensive overlay repairs on roll surfaces (Figure 22).

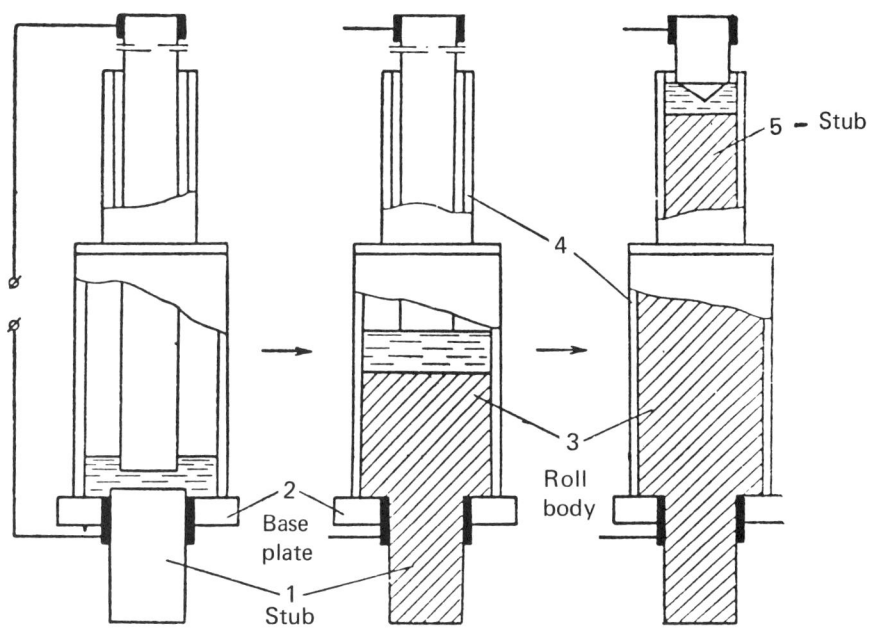

FIGURE 20 Schematic Illustration of Roll Manufacture Via ESR (from Joint Publications Research Service, 1975).

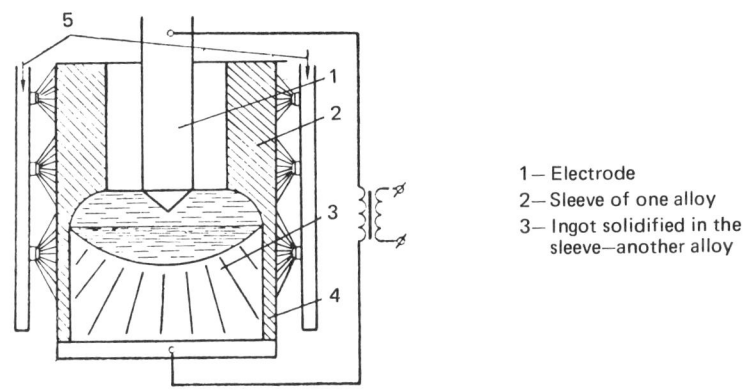

1 — Electrode
2 — Sleeve of one alloy
3 — Ingot solidified in the sleeve—another alloy

FIGURE 21 ESR Melting of Composite Metal Components (from Joint Publications Research Service, 1975).

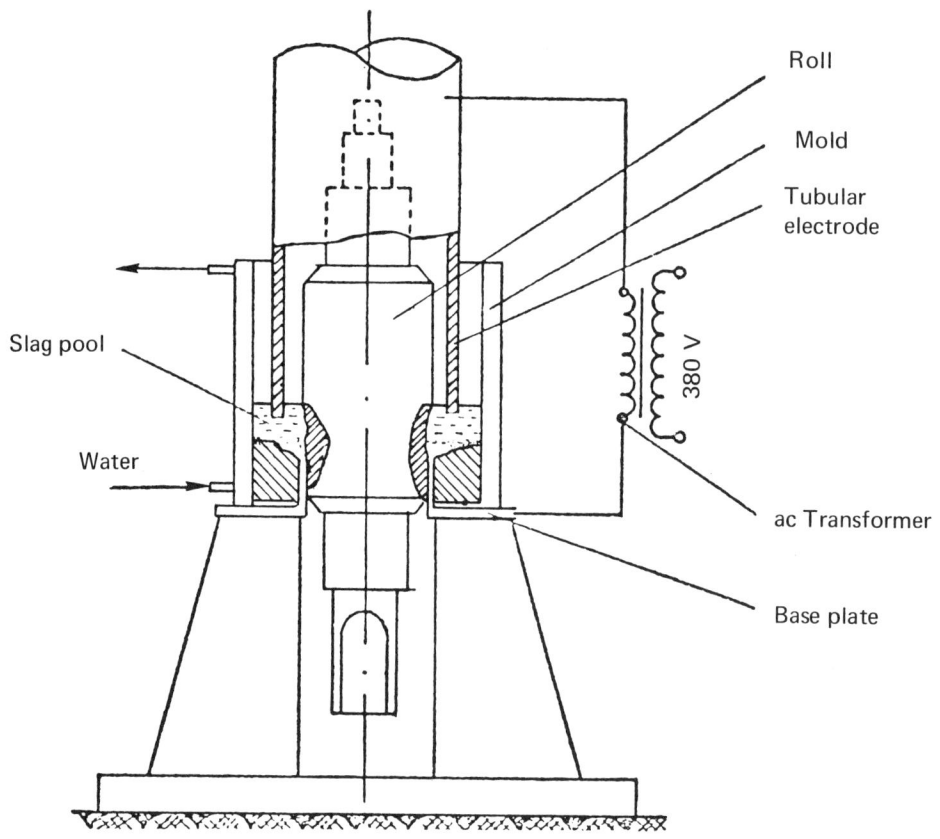

FIGURE 22 ESR Metal Overlay or Surface Repair Arrangement (from Bagshaw et al., 1973).

4. Multicontoured Castings by ESR

Multicontoured casting manufacture by ESR is gaining prominence in the USSR. The manufacture of large crankshafts, as conceived by the Soviets, has been described (Joint Publications Research Service, 1975) and is adaptable for making large generator shafts with integral flywheels or discs on the shaft. This procedure eliminates procurement of separately forged shaft and discs and, subsequently, welding them to fabricate the generator shaft.

5. Bifilar ESR System

The bifilar or paired-electrode ESR system (Figure 23) is another novel concept that has found applications in the manufacture of square and rectangular shaped ingots. The use of single-pair or multiple-paired electrodes in electrical furnaces is not a new concept. Electrodes delivering power to salt baths and to ferroalloy manufacturing furnaces normally are connected to the power source in the bifilar arrangement.

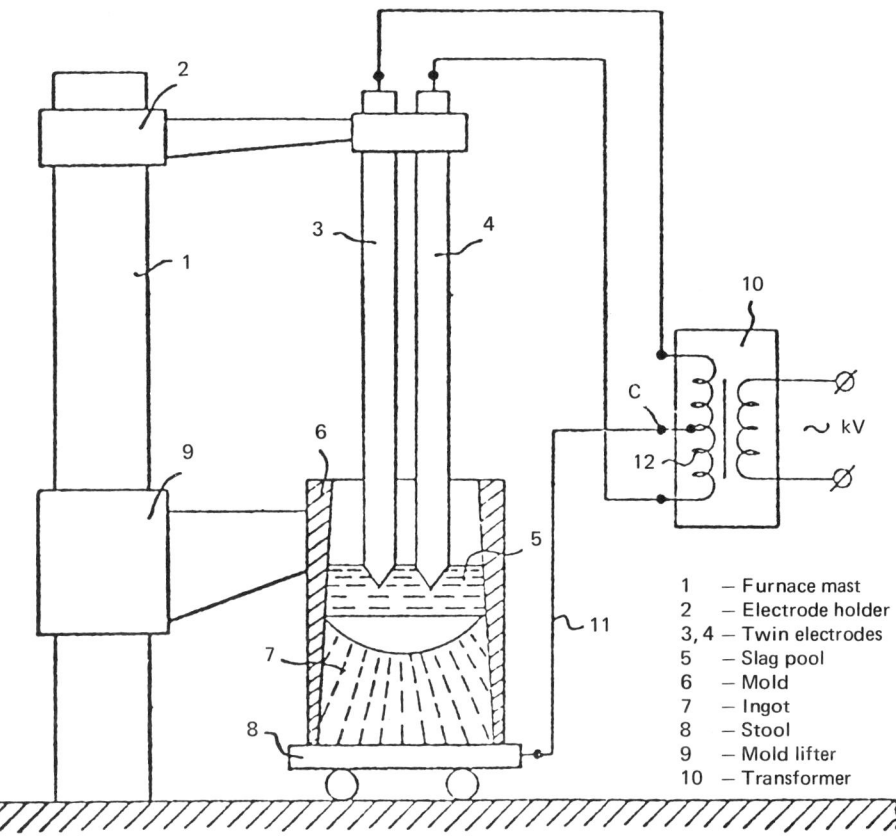

FIGURE 23 Bifilar ESR Furnace Setup for Making Square and Rectangular Ingots (from Paton et al., 1969).

This variant supplies power to the slag solely by means of the multiple electrodes. Numerous geometrical arrangements and multiple phases have been used with varying degrees of success. The method of heat generation, the resulting heat flows, and ingot structure remain essentially similar to those existing in the simpler furnace designs although the detailed temperature gradients probably differ slightly. Insufficient evidence is available on this point to allow any precise examination of the thermal differences in these methods.

The principal advantage of a bifilar system is the shortened electrical loop. The current flow in this system could pursue two parallel paths: (a) electrode-slag-electrode or (b) electrode-slag-metallic bath-slag-electrode. The magnitude of the current depends upon the resistance in each of the circuits and other factors, such as dimension of electrodes, size of the slag pool, depth of immersion of electrode in the slag pool, location of the electrodes with respect to the mold side wall, and the distance between the electrodes.

Bifilar electroslag melting regime studies have been conducted under Air Force Project 9-161 Contract AF 33(615)-5430. These studies, although preliminary in nature, have indicated that the bifilar ESR system efficiency improves if the electrical circuit is limited to electrode-slag-electrode. However, this is not always achieved in practice, and the electrodes often show a tendency to fuse-off in a nonuniform manner resulting in a nonuniform structure.

The Soviet bifilar ESR furnaces have undergone design changes since their first introduction. Several complicated changes were made in the electrical power system to correct nonuniform melting of the electrodes and excessive stirring in the metal pool leading to segregations in certain grades.

The suspension and alignment of one or more pairs of electrodes is quite time consuming. Higher productivity and lower power consumption claims made for Soviet design bifilar furnaces still must be proven in commercial practice against similar claims made for Western monofilar ESR slab casting systems.

6. Manufacture of ESR Hollow Ingots

Another adaptation of the versatile ESR process is to manufacture hollow ingots and hollow shaped castings (Figure 24). The problems relative to the casting of hollow ingots involve:
1) Extraction of the mandrel without its destruction
2) Providing adequate cooling of the mandrel to prevent its wear
3) Control of the smoothness of the inside diameter (ID) and outside diameter (OD) surfaces
4) Cost-effectiveness of the process.

An ESR hollow ingot is cooled on the outside and inside surfaces. As a result, the molten metal pool depth during the casting of the hollow ingot remains small. Average dendrite arm spacing in hollow ingots made with a water-cooled core are much smaller than those in solid ingots of equivalent diameter. The chemical and microstructural homogeneity of the hollow ingot are improved vastly. Consequently, the use of near-net-shaped ESR ingots and preforms for the manufacture of a variety of semi-finished components for aerospace and similar critical applications is advantageous from the viewpoints of material usage, reduced number of mill processing operations, mechanical design, reliable performance and overall cost reduction.

Different techniques of casting ESR hollow preforms have been discussed adequately in reports of investigations (Bhat, 1971, 1972, 1973; Norcross, 1971; Klein, 1972; Hoyle, 1973; Paton et al., 1972; Ujiie et al., 1972). In early efforts, hollow ingots were made using tapered mandrels, consumable mandrels, or collapsible mandrels. An assembly of multiple electrodes was melted in the annulus of the mold and mandrel. Melts were initiated by charging molten flux. The ingots produced were small, ranging in height from 15 to 30 inches. Satisfactory ESR hollow ingots of three grades of low-alloy high-strength steels, two grades of precipitation hardening chromium-nickel alloys and maraging 250 grade ultra-high strength steel were produced. Similarly, ESR rings 6 to 10 inches high were cast of heat-resisting alloys A-286, Inconel 600, René 41, Inconel 718, and U-700 using cast electrodes 2 inches round and 1-3/4 by 12 inches rectangular. The mechanical properties achieved in these ESR hollow ingots in the cast and wrought conditions are summarized in Tables 24 and 25.

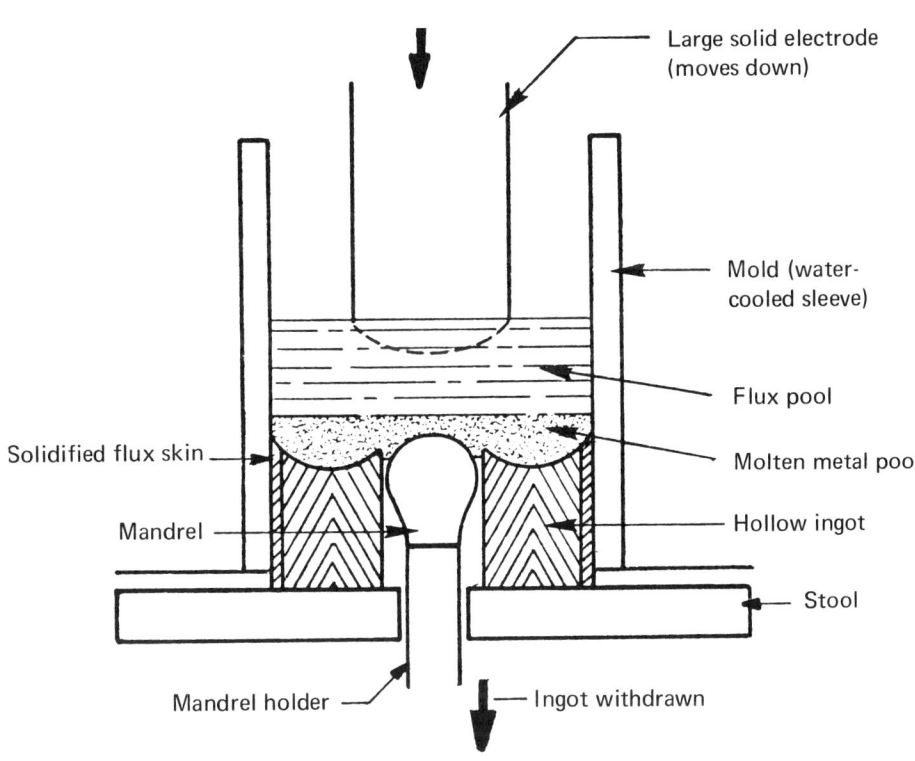

FIGURE 24 Schematic Representation of ESR Hollow Ingot Manufacturing Technique through Melt Extrusion (from Joint Publications Research Service, 1975).

The use of hollow or solid electrodes of small cross section increased the costs of ESR hollow ingots with lengths over 12 inches. Electrodes, prepared from cast slabs, are useful in producing hollow ingots of length up to 30 inches despite the higher electrode costs. The main drawbacks of the disposable mandrel are its initial cost, setup, and other costs associated with its removal from the hollow ingot. These difficulties narrowed the choices of hollow ESR ingot manufacturing systems to a short mandrel that is ejected progressively from the hollow ingot or to extruding liquid metal over a water-cooled mandrel and withdrawing a solidified hollow ingot.

TABLE 24 Mechanical Properties of ESR Cast Hollow Ingots[1] and Wrought Rings[2] of Low-Alloy Ultra-High-Strength Steels.

Material and Ingot Size	Heat Treatment	Test Direction	0.2% Yield Strength (KSI)	Tensile Strength (KSI)	Elongation (% in 2")	Reduction of Area (%)
4335 V, 3-1/2" Wall Cast Hollow	1600° F O-Q	Axial	212	240	8.0	22
	Temper--500° F	Tangential	209	238	8.5	23
4335 V, 1-3/4" Wall Wrought Ring	1600° F O-Q	Axial	215	243	10.5	30
	Temper--500° F	Tangential	211	242	11.0	29
300 M, 4" Wall Cast Hollow	1650° F Salt Quench	Axial	232	278	7.0	20
	Temper--550° F, 2 + 2	Tangential	235	282	8.0	22
300 M, 2" Wall Wrought Ring	1650° F Salt Quench	Axial	242	290	9.0	27
	Temper--550° F, 2 + 2	Tangential	245	291	9.5	28
D6, 4" Wall Cast Hollow	1650° F Salt Quench	Axial	208	248	5.5	16
	Temper--600° F	Tangential	206	247	6.5	18
D6, 1" Wall Wrought Ring	1650° F Salt Quench	Axial	211	269	7.0	20
	Temper--600° F	Tangential	210	264	6.0	19
	1100° F	Axial	172	236	14.0	30
		Tangential	170	230	13.0	28
17Cr-14Ni PH, 4" Wall Cast Hollow	1875° F Solution Treat.	Axial	170	189	15.5	38
	Age--900° F	Tangential	172	191	15.5	39
17Cr-4Ni PH, 1-3/4" Wall Wrought Hollow	1900° F Solution Treat.	Axial	173	195	19.0	42
	Age--900° F	Tangential	175	196	19.0	43
Maraging 250 KSI, 3" Wall Hollow	1500° F, 1 hour	Axial	238	244	6.0	28
	925° F, 6 hours	Tangential	242	251	7.0	30
Maraging 250 KSI, 1" Wall Wrought Ring	1500° F, 1 hour	Axial	248	258	10.0	35
	925° F, 5 hours	Tangential	251	260	11.0	38

[1] The ESR hollow ingots varied in size from 8 to 16 inches in outside diameter. Weight of the hollow preforms ranged from 220 to 462 pounds.

[2] The rings rolled from the ESR hollow preforms varied in size from 20 to 40 inches in diameter.

SOURCE: Bhat, 1973.

TABLE 25 Mechanical Properties of ESR Cast Hollow Ingots and Wrought Rings of Heat-Resistant Alloys and Superalloys.

Material, Ingot Size, and Heat Treatment	Test Direction	0.2% Offset Yield Strength (KSI)	Tensile Strength (KSI)	Elongation (%)	Reduction of Area (%)
A-286, 3-1/2" Wall Ingot, 1800° F ST, 1325° F 16 hours	Axial	100	150	16	30
	Tangential	101	151	17	31
A-286, 1" Wall Rolled Ring, 1800° F ST, 1325° F 16 hours	Axial	102	151	24	40
	Axial	103	154	25	41
Inconel 600, 4" Wall Ingot	Axial	64	102	19	39
Inconel 600, 1" Wall Rolled Ring	Axial	55	100	35	56
	Tangential	58	101	38	58
René 41, 2" Wall Rolled Ring, 1950° F 1/2 hour WQ, 1400° F 16 hours AC	Axial	124	181	19	24
	Tangential	120	178	18	22
Inco 718, 1-3/4" Wall Rolled Ring, 1800°F 1 hour + 1325° F 16 hours	Axial	161	190	25	34
	Tangential	168	198	26	35
U-700, 4" Wall Ingot, 2000° F 1 hour AC	Axial	131	165	8	17
	Tangential	128	161	8	18
U-700, 2" Wall Forged, 2000° F 1 hour AC, 1550° F 24 hours AC, 1400° F 16 hours AC	Axial	148	198	18	24
	Tangential	144	196	18	22

SOURCE: Bhat, 1973.

Two promising hollow ESR ingot manufacturing systems that have evolved from Mellon-Nutek (Bhat, 1972 and 1973) research studies are shown schematically in Figures 24 and 25. Both systems permit the use of large section electrodes. The basic arrangement, shown in Figure 24, uses a single electrode and a stationary straight-sided water-cooled sleeve mold. The basic polyelectrode system shown in Figure 25 has some versatility. A funnel-shaped water-cooled sleeve mold permits the introduction of one or more large-section electrodes.

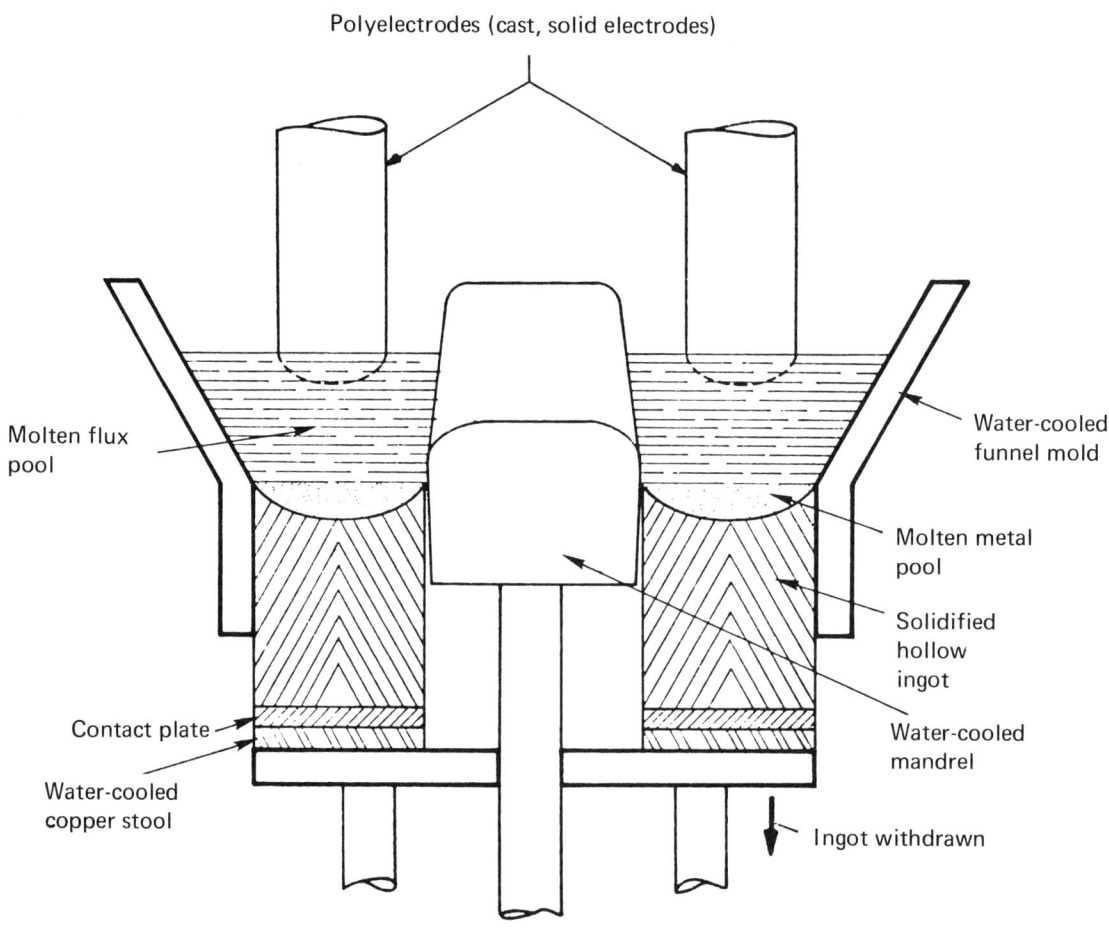

FIGURE 25 Schematic Representation of ESR Hollow Ingot Production Setup Using Funnel Mold and Polyelectrodes (from Joint Publications Research Service, 1975).

To successfully operate this ESR system, all the molten metal must be confined within the straight section of the funnel mold. The mandrel can be fixed so that its top protrudes above the flux level or just above the molten metal level. The withdrawal of the ingot creates an action of squeezing the molten metal in the annulus of the mold and mandrel. The outer diameter and inner diameter shape and wall thickness of the hollow ingot can be controlled by fitting a collar and mandrel having the desired contours machined respectively on the inner diameter and outer diameter surfaces. This type of ESR casting can be described as an "extrusion" or squeeze melt process.

At present, the Air Force Materials Laboratory (AFSC) at Wright-Patterson Air Force Base is sponsoring a program aimed at development of a viable production method for the manufacture of ESR superalloy hollows. The anticipated cost savings resulting from starting with a hollow instead of a solid justifies the program and ingot cost savings, on the order of 30 to 70 percent, are projected for the aircraft turbine components chosen in the contract. While the alloys in this contract are superalloys, René 95 and Astroloy, the process to be developed is expected to be applicable to other alloy systems. Other applications may include preforms for gun tubes and hollow billets for use as starting stock in the production of seamless tubing for corrosion and nuclear applications.

The program initially investigated, on a subscale basis, defined a number of different techniques that could be used for producing ESR hollows. A technique – hot piercing – was chosen to be the most promising method from the technical and cost standpoint. This process is being scaled up to produce ingots up to 20 inches outer diameter by 10-1/2 inches inner diameter.

7. Manufacture of Small and Large Pressure Vessels

The essential features of an ESR setup for manufacturing pressure vessels with integral dome and ports is shown schematically in Figure 26. The mold is made in several sections with each segment assembled on a central water-cooled post. The multiple-electrode suspension assembly is designed in such a way that the position of each electrode can be changed as desired. Molten flux is first delivered to initiate the melt and the dome is cast first. Then the inner water-cooled mold section conforming to the contour of the vessel is lowered around the mandrel and the casting is built up in successive segments. A similar Soviet concept for building pressure vessels with integral ports has been described (Paton et al., 1972).

8. Manufacture of ESR Castings by the YOZO Technique

The YOZO technique (Ujiie et al., 1972) is essentially an ESR weld-forming method specially developed for the manufacture of castings for the petrochemical and nuclear industries in Japan. This technique, adapted by Mitsubishi Heavy Industries, Ltd., is claimed to produce castings superior in metallurgical characteristics to those produced through the centrifugal casting method.

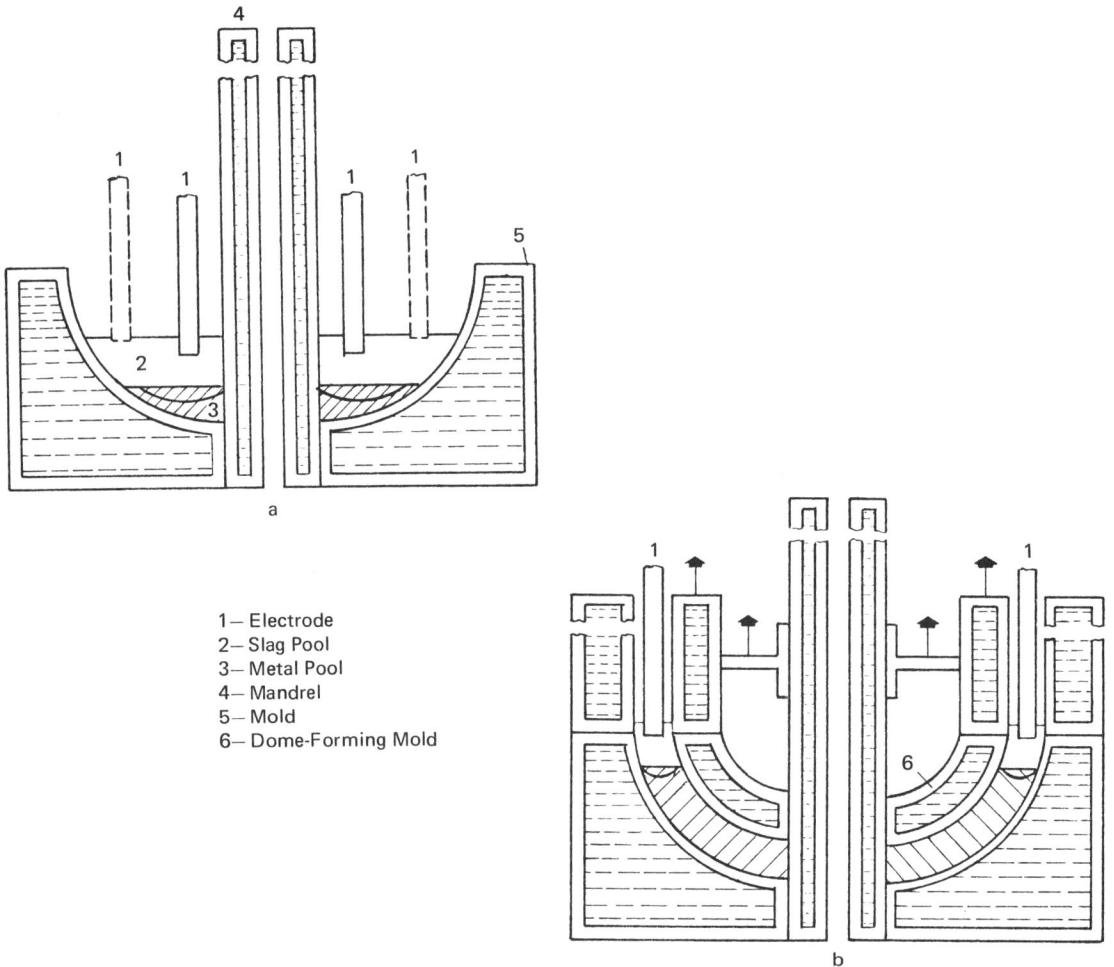

1— Electrode
2— Slag Pool
3— Metal Pool
4— Mandrel
5— Mold
6— Dome-Forming Mold

FIGURE 26 Essential Features of an ESR Setup for Manufacturing Pressure Vessels with Integral Dome (from Joint Publications Research Service, 1975).

The YOZO method of casting straight tubes is shown in Figure 27. Methods of forming more complicated castings have been reported in the literature (Joint Publications Research Service, 1975). The principal disadvantages of the technique are the production of thin electrodes and the repetitive process setup labor costs.

9. <u>Manufacture of Ingots Using Metal Composite Electrodes</u>

Composite electrodes are made either by filling a tube with alloying constituents or by assimilating scrap in the required form. The alloy-filled tube electrodes were used in an earlier ESR system (Hopkins process) and the latter system is used at the Riken Piston Company in Japan (Madano, 1969). The alloy-filled tube system was moderately successful, but the occasional presence of unmelted alloying constituents in the ESR ingot was a problem. In the assimilated scrap system, on the other hand, the composite electrode was swaged prior to remelting, and the use of very thin section electrodes to form a comparatively large-cross-section ingot provided a homogeneous ESR ingot in the "scrap" system.

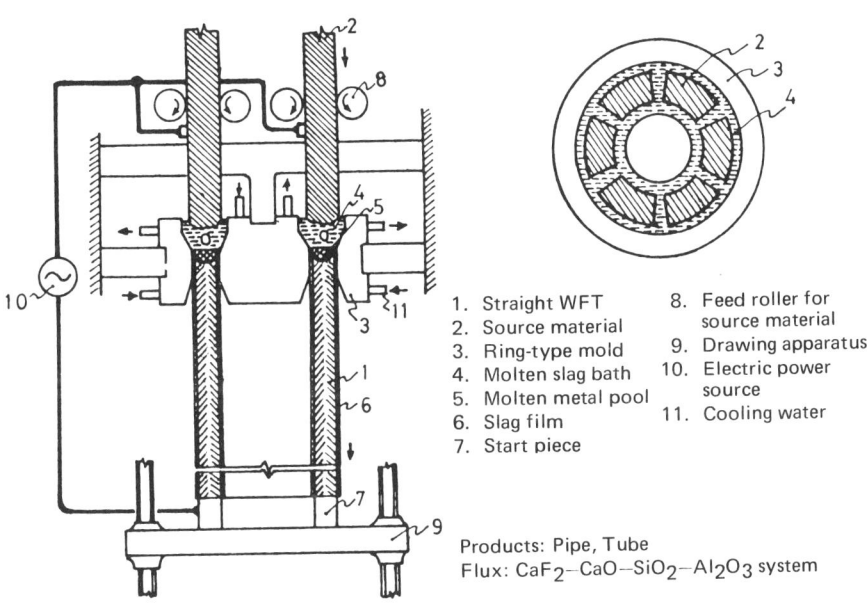

1. Straight WFT
2. Source material
3. Ring-type mold
4. Molten slag bath
5. Molten metal pool
6. Slag film
7. Start piece
8. Feed roller for source material
9. Drawing apparatus
10. Electric power source
11. Cooling water

Products: Pipe, Tube
Flux: CaF_2–CaO–SiO_2–Al_2O_3 system

FIGURE 27 Diagram of Electroslag Melting of Tube Billets by the YOZO Method (from Joint Publications Research Service, 1975).

10. Manufacture of Ingots or Castings Using Molten Metal Feed, Scrap Feed, or Powdered Metallic Component Feed

ESR equipment that has been used to manufacture ingots or castings using either molten metal, scrap, or powder metallic component feed, is similar (Bhat, 1973). Melting is done in a funnel- or T-shaped mold. The slag is kept molten by feeding electrical power through either a consumable or a nonconsumable electrode. Molten metal may be fed through a hollow graphite electrode that also serves as a current conductor to keep the slag molten. The process is initiated by charging molten slag into the mold.

The basic features of an ESR setup to recycle scrap are shown schematically in Figure 28. The mold has a baffle or a dam to prevent transfer of any unfused metallic particles into the solidifying pool of metal in the narrow section of the mold. Such systems are promising for the direct manufacture of very large ingots or castings, bypassing the use of solid consumable electrodes. The nonconsumable electrode, shown in Figure 28, could be either a solid conductor or a plasma.

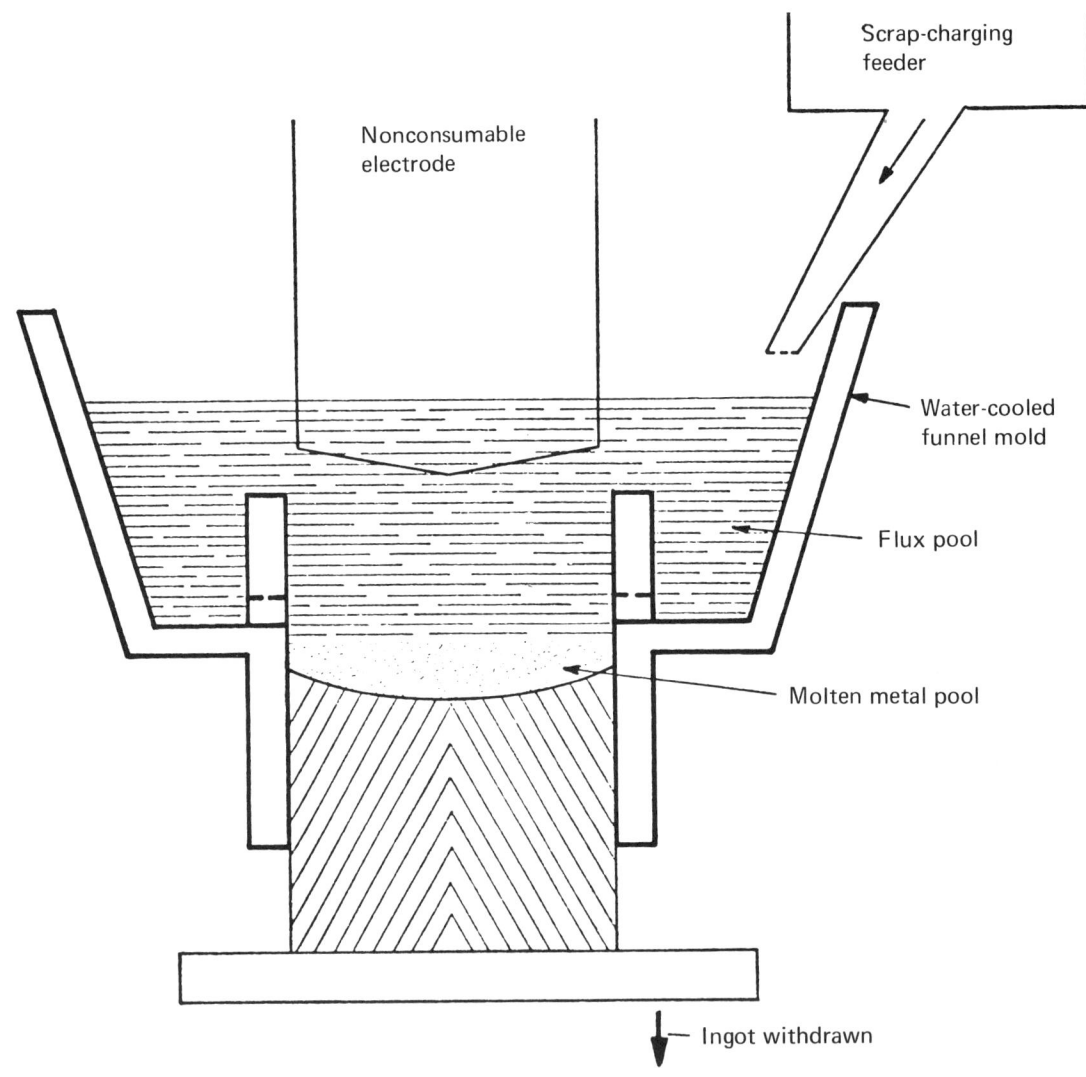

FIGURE 28 Schematic Arrangement of Electroslag System for Scrap Consolidation (from Bhat, 1973).

11. Direct Electroslag Melting

The direct electroslag melting process envisages conversion of sponge iron or prereduced pellets and alloying materials into homogeneous metal ingots. The process is operated using electrical power from either single-phase or three-phase power sources, and the equipment is similar to that shown in Figure 28. Potential industrial applications of the direct electroslag melting process are summarized in Table 26.

TABLE 26 Direct Electroslag Melting.

	Mild Steel or Low-Alloy Steel	Alloyed Steel	High-Alloy Steel	Ferro-Alloy Master Alloy
Charge	Iron Sponge Ferro-Alloys	Iron Sponge Master Alloy	Billets of Mild Steel or High Carbon Steel Master Alloy	Oxides Reduction Refining
Electrode	Nonconsumable Graphite	Nonconsumable	Consumable	Nonconsumable

SOURCE: Thomas, 1971.

In spite of considerable efforts, including some production experience, processes involving a substantial proportion of the feed as particulate material always have encountered difficulties.

C. Conclusions and Recommendations

1. Conclusions

    a. No novel adaptations of the electroslag remelting process have reached commercial status in the United States, excluding slab ingots.

    b. Production of certain types of casting via ESR reportedly has reached industrial acceptance in other countries. These castings include roll bodies, clad rolls, large rings, valves, crankshafts, tubes, and similar shaped castings.

    c. Characterization data are lacking on cast material produced through various adaptations of the ESR process.

d. Many novel ESR concepts are in developmental stages. Reports on the subject do not provide specific reasons for consideration of the ESR casting process although general statements extoll the virtues of a particular idea.

2. Recommendations

a. As-cast ESR metal should be characterized.

b. Various metallurgical problems including mechanical properties at the joints made by electroslag casting should be evaluated.

c. Equipment designs for the manufacture of multiple small-cross-section ESR ingots should be refined.

d. ESR casting of large ingots and large shaped castings using a combination of solid and molten metal electrodes should be evaluated.

e. Composite shapes produced by the ESR process should be exploited.

f. ESR hollow ingot development programs should be continued.

g. Each novel ESR application should be evaluated from a systems viewpoint, based on the specific requirements of the end product.

VII. ELECTROSLAG EQUIPMENT TECHNOLOGY

A. Current Equipment Technology

1. Description

ESR equipment technology has evolved differently in various parts of the world as process and equipment developers attempted to optimize process parameters of the relatively new technology while seeking to maximize the electrical and mechanical efficiencies of ESR furnaces. This effort continued under the added pressure of reducing equipment and operating costs and the desire for maximum ingot productivity. As a consequence, the variety of ESR furnace designs in operation today resulted from the demands of the users and the various technological approaches of the equipment manufacturers.

Excluding for the moment the infinite variety of possible mechanical variations in furnace design, the following basic types of electrical system (Figure 29) are in use today:

1) DC with electrode negative or positive
2) AC single-phase line frequency, coaxial and noncoaxial
3) AC low frequency (3 to 10 Hz)
4) AC three-phase; three of six electrodes
5) AC bifilar; two, four, or six electrodes

The various ESR mold configurations in use today are summarized as follows:

1) Full length, fixed molds
   a) Round – up to 60 inches in diameter
   b) Square
   c) Rectangular (slabs) – 30 by 80 inches in the USA
2) Conventional collar molds – moving mold or ingot withdrawal
   a) Round – up to 92 inches diameter (maximum size at Roechling)
   b) Square
   c) Rectangular (slab) – 10 by 30 inches in production and larger ones under development

3) Tulip mold – a variation of the collar mold, is presumably in limited use within the USSR for the production of slabs, large round ingots, and multiple ingots (see Figures 15 and 18). Basically, a tulip mold is a water-cooled copper mold configuration consisting of an upper expanded portion (the region that contains the molten slag and into which the electrode or electrodes are immersed) and a reduced lower section in which the slag-metal interface is maintained and where the ingot solidifies and is withdrawn continuously.

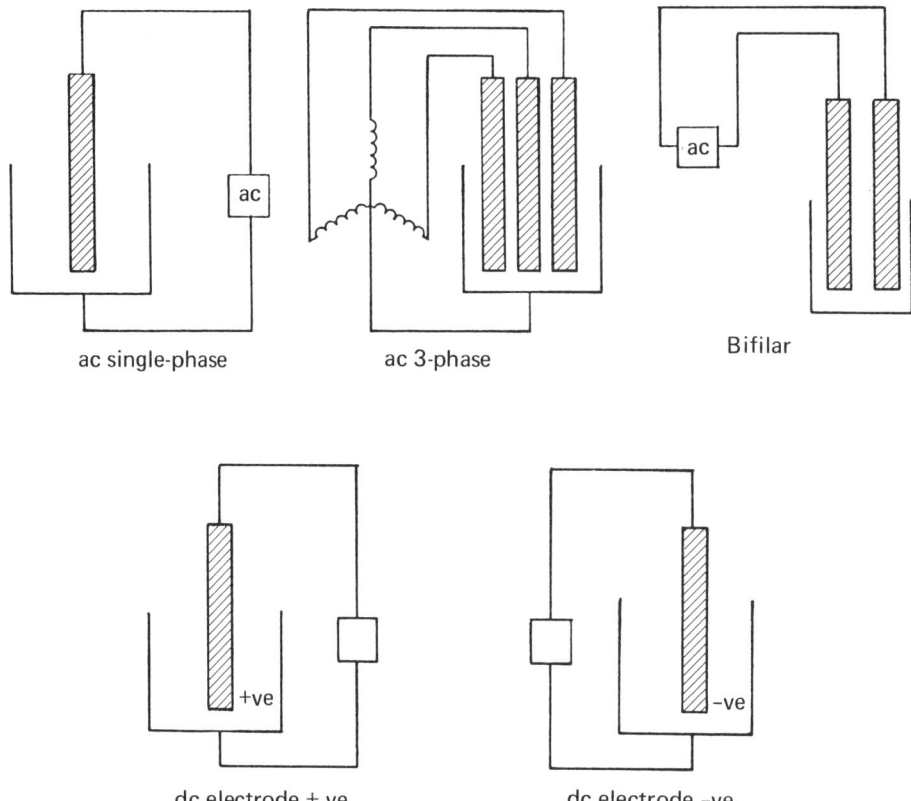

FIGURE 29 Principal Ways of Supplying Power for the Electroslag Process (from Duckworth and Hoyle, 1969).

Electrode configuration and handling vary with each furnace system and intended application. Electrode cross-sectional area relative to the ingot cross section also is variable. Systems in common use are:
1) DC, AC single-phase line frequency, and low-frequency powered furnaces usually operate with one electrode at a time melting into a larger diameter ingot mold.

a) One electrode into one fixed full length mold.
b) One electrode into one collar mold, moving mold or ingot withdrawal.
c) Multiple electrode operation in which a series of electrodes are melted in sequence, known as the electrode change technique; either into a fixed full length mold or preferably into collar molds.
d) An exception to the above is the Roechling furnace that can operate with one, two, or four changeable electrodes each individually driven and separately powered by AC low frequency (3 to 10 Hz) power supplies.
2) Three-phase AC furnaces use three electrodes.
3) Bifilar furnaces operate with electrodes in pairs.

Figures 30 and 31 illustrate typical recent ESR furnace designs. A survey of ESR installations is reported in Appendix B.

2. Power Supplies for ESR Furnaces

Electroslag remelting is a resistance heating process and hence its low voltage, high current power requirement is not particularly conducive to highly efficient power transmission. Required process voltages are from 15 to 70 V across the slag with currents varying from 500 to 100,000 A, depending on ingot cross-sectional area. Distribution voltages are usually from 480 V to 14 kV, requiring transformer equipment.

Developments in process and operation of electroslag refining require infinitely variable voltage and current control over a wide range with an accuracy of ± 0.5 percent. Saturable reactors or thyristors in combination with closed-loop regulation systems controlling the supply to a normal ESR transformer meet this requirement admirably, and an increasing proportion of new installations have this facility. Stepless variable output transformers with a continuously tapped secondary winding also may be used but they do not respond as quickly to load variations or control signals. Some type of transformer secondary current limiting usually is designed into the power supplies to protect against complete short circuits if they accidentally occur.

Power supply cables should have the shortest possible route from transformer to electrode clamp and be sited on the same route for base-plate return to the transformer to reduce the inductive loop and maintain transmission efficiency. Power factors of various single-phase line frequency furnace designs and capacities can vary from 0.35 to 0.92. Capacitance may be used to reduce the primary kilovolt ampere maximum demand charge by raising this power factor under full load approximately to 0.90, but any economic advantage to be gained varies according to furnace and local site conditions.

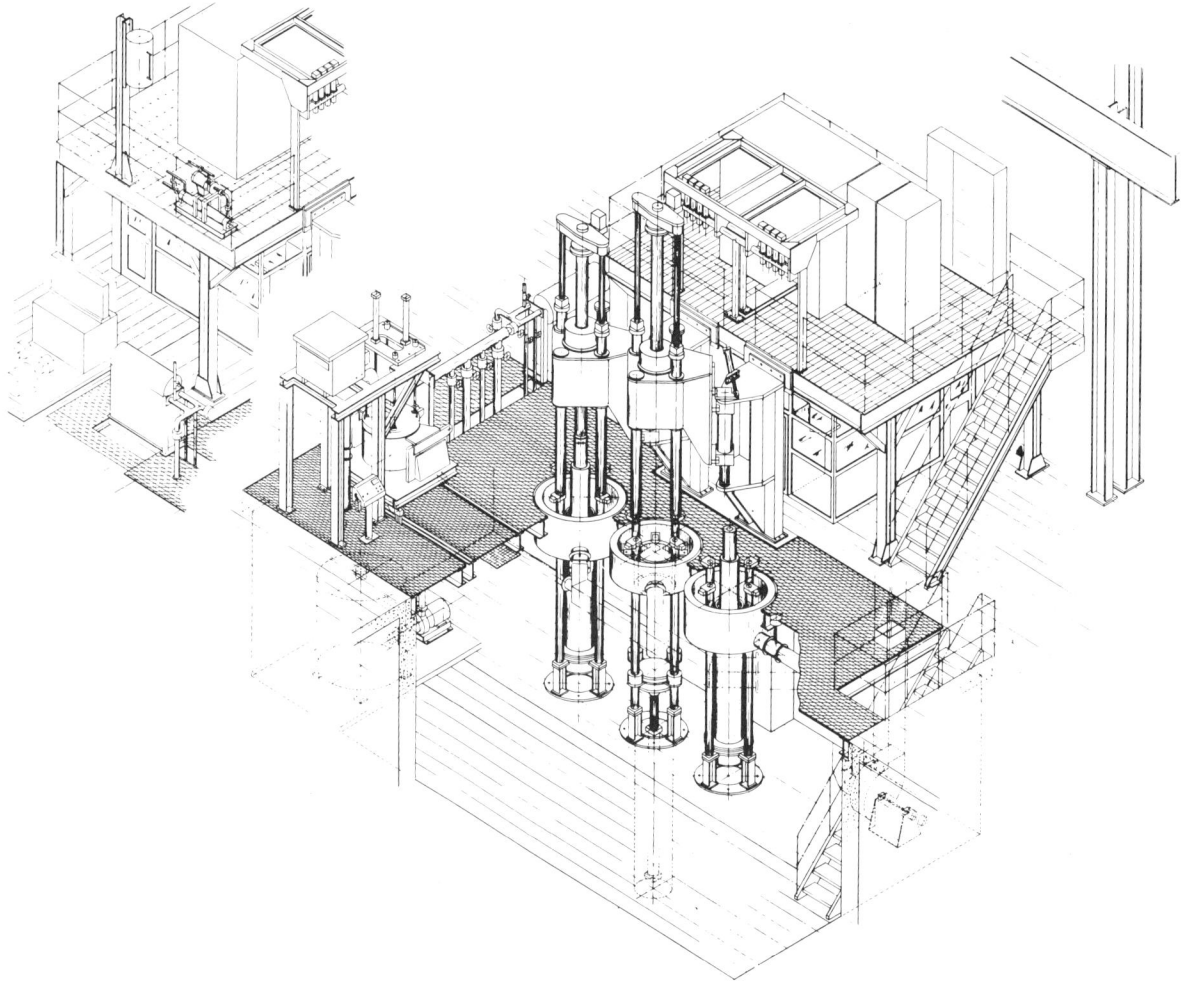

FIGURE 30 Three-Station ESR Furnace for 20- to 40-Inch Diameter Ingot (from CONSARC Corporation).

This furnace has fixed and ingot withdrawal capability, four-conductor return bus system, sliding contact to minimize stray magnetic field and to reduce the inherent reactance of the furnace, two-electrode suspension-drive systems, and electrode change capability.

The majority of ESR furnaces in operation today are single-phase AC (coaxial and noncoaxial) with a trend towards low frequency AC for furnaces larger than 30-inch diameter ingot capability, followed by AC bifilar, then AC three-phase, DC positive, and DC negative.

FIGURE 31   Lukens Steel ESR Slab Furnace of CONSARC Design (from Lukens Steel Company).

Single-phase AC, 60 Hz, 60,000 amperes, 30 by 80 inch slab ingots.

The relative merits of the five basic power supplies that can be used for electroslag melting are compared in Table 27 (Luchok and Wooding, 1974). A comparison of slab ingot production ESR furnace characteristics when using AC single-phase, bifilar, and three-phase furnaces is presented in Table 28 (Luchok and Wooding, 1974).

TABLE 27 Comparison of Different ESR Power Modes.

	AC (1 ∅)	AC (3 ∅)	DC (+)	DC (−)	AC/DC (5 Hz)
Power Cost per Ton	good	good	poor	good	fair/good
Power Supply Capital Cost − 90% P.F.	good	good	poor	good	poor
Electrode Cost	good	poor	good	good	good
Furnace Cost	good	poor	good	good	good
Melt Set-up Cost	good	poor	good	good	good
Desulfurization	good	good	good	poor	good
Pool Depth	good	good	poor	good	good
Absence of Inherent Stirring	good	poor	good	good	good
Slab Melting	good	poor	good	good	good
Molten Slag Starting	fair/good	good	good	fair	good
Melt Mode Flexibility	poor	poor	fair	fair	good
Supply Line Balance	poor	good	good	good	good
Supply Line Flicker	good	good	good	good	poor
Power Supply Reliability (mean time between failures)	good	good	fair	fair	poor

Basis for Comparisons: Coaxial furnace design for modes AC (1 ∅), DC (+), DC (−), and AC/DC (5 Hz); high fill ratio (electrode cross sectional area 60 to 85 percent of ingot cross section); melting conducted in open crucible (no inert gas); ingots over 12 inches in diameter (or equivalent slab section); and power supply with stepless control of current and voltage.

SOURCE: Luchok and Wooding, 1974.

Obviously, it is not valid to compare melt modes simply on the basis of the number of "good" or "poor" ratings they received since these indices are not weighted equally for all situations (Roberts, 1970). For example, for an ingot producer contemplating the installation of a rather large electroslag furnace on a low capacity power distribution system, the problem of supply line

TABLE 28 Comparison of Slab Ingot Production ESR Furnace Characteristics.

	Single-Phase Coaxial Furnace	Bifilar Furnace	Three-Phase Furnace
Number of Electrodes (without electrode change)	One.	Two.	Three.
Electrode Cross-Section	Large, approximately 70% cross-section of ESR ingot.	Smaller; two electrodes required.	Smaller; three electrodes required.
Cost of Making Electrodes	Minimum.	High. Electrodes must be very long relative to cross-sectioned area.	Higher than for single-phase or bifilar.
Cost of Electrode to Stub Welds	Minimum (only one stub to be aligned and welded per melt).	Higher (two stubs to be aligned and welded per melt).	Higher (three stubs to be aligned and welded per melt).
Electrode Change Possible?*	Yes.	Yes, but more difficult.	Yes, but more difficult.
Electrode Change Complexity*	Minimum.	More complex. Two electrodes. Electrode alignment more critical.	Complex. Three electrodes, three drive systems. Electrode alignment most critical.
Pool Depth for Given Melt Rate	Moderate.	Much deeper.	Much deeper.
Stirring	Minimum.	Vigorous.	Vigorous.
Segregation Tendencies	Minimum (alloy dependent).	Greater (alloy dependent).	Greater (alloy dependent).
Average Melt Rate Possible for Given Pool Depth	At least 30% greater than three-phase or bifilar.	Less than single-phase because of excessive pool depths.	Less than single-phase because of excessive pool depths.
Furnace Productivity (tons per year)	Approximately 30% greater than three-phase or bifilar.	Less than single-phase.	Less than single-phase.
Hot-Topping: Loss of Productivity	Minimum.	Greater, more time and more difficult to hot-top.	Greater, more time and more difficult to hot-top.
Hot-Topping: Loss of Product Yield	Minimum.	Greater, more difficult to hot-top.	Greater, more difficult to hot-top.
Mold Life for Given Melt Rate	Better.	Poorer (added heat flux to mold wall because of vigorous slag stirring; also risk of parasitic arcing).	Poorer (heat flux to mold wall greater because of vigorous slag stirring).
Ingot Compositional Control and Uniformity	No problem.	More difficult due to pool agitation.	More difficult due to pool agitation.
Danger of Slag Entrapment	No problem over wide range of melt rate.	Can be a problem at lower melt rates.	Can be a problem at lower melt rates.
Interaction of Slag with Atmosphere	Minimum.	Greater due to excessive stirring.	Greater due to excessive stirring.
Electrode Drive System	Minimum complexity; high system reliability.	More complex.	More complex, three drives; lower system reliability.
Potential for Hydrogen Pick-up	Minimum.	Greater because of vigorous slag and metal stirring.	Greater because of vigorous slag and metal stirring.
Three-Phase Line Balance	Not inherent. Balancer must be used if three-phase balance is required.	Not inherent. Balancer must be used if three-phase balance is required.	Inherent.

* Production of slab ingots from consecutively changed electrodes is not recommended.

SOURCE: Luchok and Wooding, 1974.

balance probably is one of his primary considerations in making the power supply choice. Another producer, however, installing the same furnace on a much larger distribution system, would not have to concern himself at all about supply line balance and would be free to make his decision based on other parameters. Similar arguments are applicable to most of the other indices used for a general comparison. Therefore, it is impossible to arrive at a universal solution concerning the best power supply in all cases.

For the future, the continuing popularity of AC single-phase melting is foreseen, with AC low-frequency being seriously considered in an increasing number of large installations.

3. Continuous Electroslag Powder Melting (CESPM)

In the CESPM process a strip electrode is fed continuously into a slag bath with the powders held magnetically to the strip until they melt, mix, and are refined in passing through the bath. An ingot is formed in a water-cooled copper collar mold and is withdrawn as fast as it is produced. A variety of alloys have been made by the process in furnaces designed by Arcos and Electrotherm SA in Belgium. A billet furnace with a 6 by 6 inch ingot section capacity has been described (Dorschu, 1973). The principle of the process is shown in Figures 32 and 33, and some operational considerations have been discussed in section V.E.3.

A CESPM unit of this design operated for several years at the former Cybermetals Corporation in New Jersey but is no longer in operation. The only other CESPM installation known to be in routine operation is used for the production of forging ingots at Cockerill-Ougree in Belgium (Descamps and Etienne, 1973). A schematic diagram of that system is shown in Figure 34. The furnace can produce ingots 40 inches in diameter and 160 inches long and can operate in the single-phase, one electrode mode or three-phase, three electrode mode depending on ingot size requirements.

From an electrical-mechanical standpoint, both of the mentioned furnaces function satisfactorily; however, insufficient data have been published by which to judge the metallurgical merits of the CESPM process (see section V.E.3).

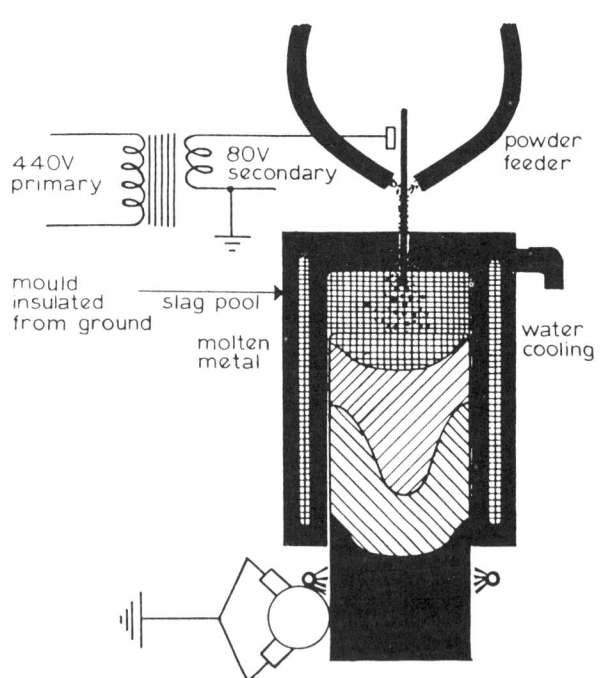

FIGURE 32 Schematic Diagram of the CESPM Process (Dorschu, 1973).

4. ESR Molds

Except for the early ESR work performed on molds in the 1930s (Hopkins, 1940), much of the later ESR technology for conventional round molds was based on vacuum arc furnace mold designs. However, it quickly became apparent that ESR molds are subjected to more severe operating conditions than equivalent VAR ones because of the resistively heated slag layer ranging from a few inches to more than 10 inches in thickness for ingots very large in diameter.

To cope with the added heat flux and thereby extend mold life, the cooling-water velocity at the mold wall was increased and the mold wall thickened. Additionally, water conditioning was required to minimize the formation of mineral scale deposits on the mold wall. If the mineral deposits that form are not removed frequently, premature mold failure results.

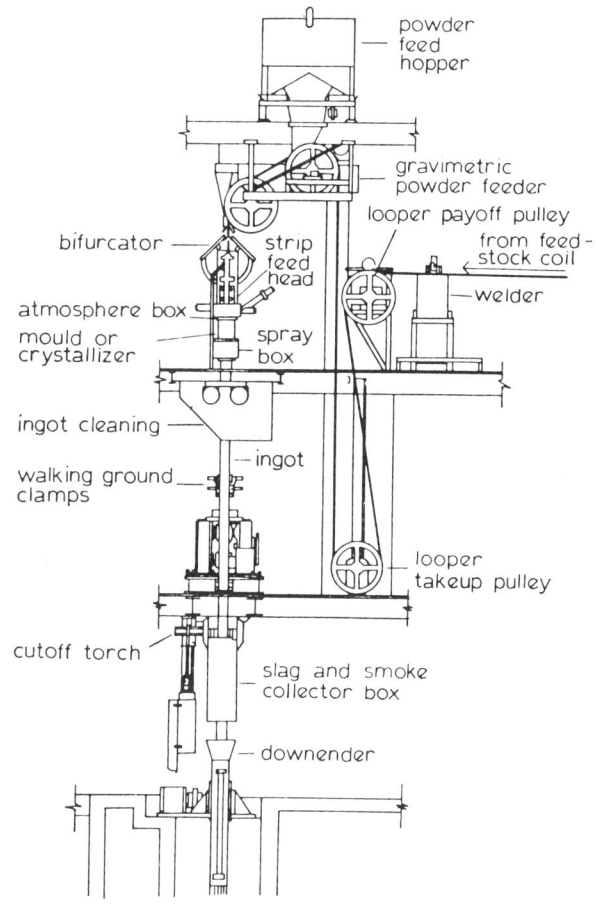

FIGURE 33 Continuous ESM Furnace and Caster (Dorschu, 1973).

While there has been much experimentation with various mold materials, at present practically 99 percent of all known ESR furnaces operate regularly with water-cooled copper molds. Although some round and square steel molds are still in use, they are short-lived and are at an economic disadvantage to copper molds.

Since the ESR process lends itself to the production of a wide variety of ingot shapes (e.g., squares and slabs), such special shapes have provided the greatest challenge to the mold manufacturers. In slab applications, the mold performance may spell the difference between economic success and failure for a given alloy product. The larger slab molds (in excess of 30 inches wide) are the most challenging. Shape retention over an extended number of heats remains a nagging problem.

Until recently, two basic design approaches existed – one is an all-welded construction in which maximum restraint through heavy bracing is applied (Cremisio and Zak, 1973) and the other (by CONSARC) is a "book mold" design in which the sides and ends are jacketed separately and held together in a manner that allows the mold sections to expand and contract in a controlled fashion, thus minimizing the tendency for permanent distortion. A more novel approach to slab mold design uses the composite mold that consists of a group of individually water-cooled copper modules (extrusions) held together in a light frame. Each module is held so that it is free to grow vertically during ESR operation with sufficient space between each module to permit them to expand laterally without disturbing the overall mold wall structure. The space between modules is too small for molten slag to enter.

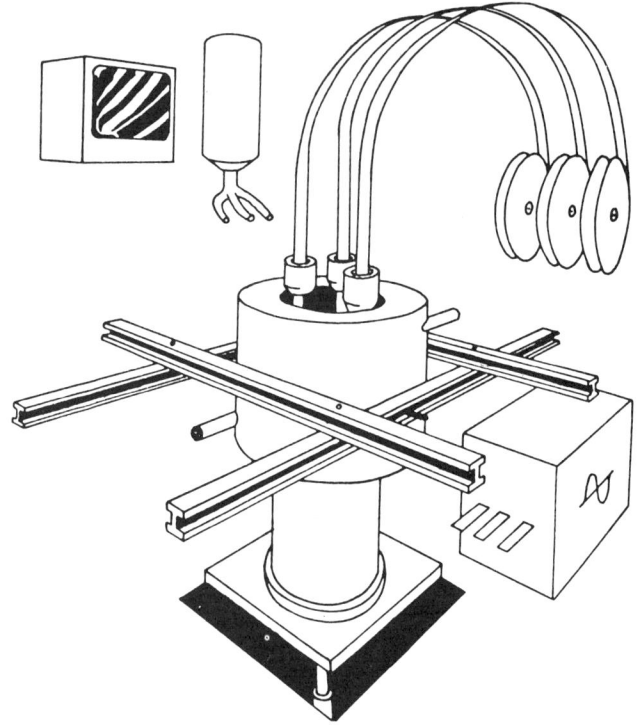

FIGURE 34 Schematic Diagram of a Three-Phase CESPM Unit for Large Ingots (Descamps and Etienne, 1973).

B. Trends and Forecast of Developments

1. High Productivity Small Ingot ESR Furnace Development

A technique for manufacturing high-integrity, small-diameter (7 inches or less) ingots of specialty steels at a volume and cost competitive with, or superior to, the conventional production of an equivalent bar or billet is needed. Such small-diameter ingots could be processed directly to final mill form by rolling or extrusion, and also could substitute for equivalent size wrought billet multiples for direct forging into parts.

Such a process is under development and has been described (Luchok and Wooding, 1974). In summary, the process consists of ESR melting in a vessel having water-cooled copper walls and a refractory bottom. A metal pool several inches deep is formed to cover the refractory bottom while the reactive slag remains in contact with the cooled copper walls. The pool is fed by one large electrode. A plurality of small-diameter ingots are withdrawn continuously from the molten metal pool (Figure 19) at a rate compatible with the formation of a sound ingot structure. It is essentially a slow continuous casting process. The process also would work with a liquid metal feed from a holding furnace and a nonconsumable electrode to maintain the slag temperature.

The commercial introduction of the process must await completion of a production-scale development program. There have been other approaches to making small ingots from large electrodes but the melting of one electrode into one ingot of smaller diameter has productivity limitations.

In the USSR, a process is under development for producing four small ingots from four electrodes using bifilar circuitry and a common slag pool with the slag metal interface in the reduced ingot forming section. Controlling the process, maintaining the position of the slag metal interface, and maintaining uniform electrode burn-off are the major problems.

2. High Aspect Ratio ESR Slab Furnaces

Some steelmakers have a substantial interest in an ESR process for producing stainless steel and nickel-base alloys in 4 to 6 inch thick slab ingot form 60 inches wide and 200 inches in minimum length.

This is a challenging problem for equipment producers, particularly in the area of mold design to meet the slab length requirement. It is equally challenging for the steelmaker to manufacture, at a reasonable cost, the thin high aspect ratio electrodes required for such a process.

The Soviets favor the ingot withdrawal approach using tulip-shaped molds. The other approach is using a conventional full-length mold. Both approaches are fraught with difficulties.

3. Roll Reclamation ESR Furnaces

Reclamation of used rolls by direct remelting in an ESR furnace is an established manufacturing technique and is seeing increasingly wider use. Also, the need exists for equipment and process know-how to resurface worn rolls by ESR cladding. Such work has been reported from a variety of sources. However, in Western countries, work on such applications is still experimental. A substantial effort to develop ESR-processed cast clad roll is under way in the United Kingdom and in Western Europe (Bagshaw et al., 1973).

4. ESR Hollow Ingot Furnace Development

Hollow ingots produced by ESR are needed for a wide range of potential applications (e.g., ring blanks, large-diameter seamless tubing, and gun barrels) providing such material is cost-competitive and technically equivalent, or superior, to conventionally produced hollow shapes. The Soviets currently appear to have a technological lead in the development of hollow ESR ingot making apparatus and processes although they are limited to modest ingot lengths and internal diameters.

A current Air Force-sponsored program is aimed at the development of a hollow ingot manufacturing process for superalloys. The program covers the development of the hollow ingot making apparatus, all process parameters, and the required process control features. The program is being conducted on a scale sufficient to permit its rapid implementation as a production process should technical and economic factors prove favorable. In Japan, a semicontinuous ESR furnace system for producing modest-diameter, light-wall cast tubing has been developed and placed into operation (Figure 27).

5. Protection Against Hydrogen Contamination During ESR Processing

As alloy steel ingots increase in size, hydrogen contamination becomes a serious problem and its content in the material must be controlled reliably within tolerable limits. Various controls of the partial pressure of water vapor in the atmosphere above the melt have been proposed. Production-scale development work is being conducted but the results either are incomplete or have received only limited publication. Hydrogen control clearly must continue to be given high priority if large ESR ingots are to be successfully produced.

Equipment manufacturers are exploring closed furnace systems in which a dry atmosphere of air or inert gases can be maintained reliably. Further work also must be done to control other sources of hydrogen in the melting system, including fluxes, starting techniques, additions, and electrodes.

Equipment for maintaining proper furnace atmospheres is available and can be adapted readily to single-electrode furnaces, but it is more difficult to provide continuous atmosphere protection for furnaces that use multiple electrodes and electrode change. Until development and operating data from large ESR systems become available, the degree of integrity required of a furnace enclosure to ensure against hydrogen contamination by the atmosphere remains unknown.

The Soviets have proposed and patented the idea of bubbling gases and other compounds through the slag bath as a means of reducing hydrogen levels during ESR processing. Success has been reported in lowering hydrogen levels of ESR processed material by application of a DC bias to AC melting (Holzgruber et al., 1969). Neither process, however, has been observed to work in commercial practice.

6. Continuous Liquid Metal ESR Processing

Liquid metal ESR processing, a nonconsumable electrode technique in which the feed metal is introduced to the system in molten form, is under study on a laboratory scale but has not been reduced to commercial practice yet. Several basic technical and economic problems must be solved before such ESR processing becomes commercial.

The rate at which an ESR ingot solidifies is limited by the structural requirements of the ingot. In current ESR practice, this rate may vary from 250 pounds per hour for an ingot 6 inches in diameter to slightly over one ton per hour for ingots 40 inches in diameter, depending on alloy grade and specification requirements. Therefore, the feed metal must reside in some stable form – either as a solid electrode, as practiced today, or as liquid metal in a holding furnace. The choice ultimately is one of economics: Is it cheaper to make and inventory solid electrodes, with the operational flexibility it provides, or to hold molten metal for extended periods of time and feed it into a mold?

The liquid feed approach has the greatest appeal for applications involving large process runs of relatively few alloys that have an ingot structural tolerance for high solidification rates or where the cost of manufacturing conventional electrodes, perhaps because of the shape requirement, becomes prohibitive. From an equipment standpoint, the most formidable problem is providing a durable and reliable nonconsumable electrode. Water-cooled copper electrodes have been tried with only minimal success. Graphite electrodes work, up to a point, but would be a source of carbon contamination for certain alloys. Theoretically, a plasma torch appears to be best suited as a noncontaminating, nonconsumable electrode heat source.

7. Large Ingot ESR Furnaces Greater than 60 inches in Diameter

A large Leybold-Heraeus designed ESR furnace is now operating that can produce 160-ton ingots 92 inches in diameter (Wahlster and Schumann, 1973). Full-length ingots up to 60 inches in diameter or relatively short ingots 88 inches long and 92 inches in diameter, have been produced successfully. However, technical difficulties were encountered when attempting to produce full-length, larger-diameter ingots. These difficulties have not been publicized and Leybold-Heraeus concludes that their proposed furnace concept shows very promising prospects for the future development of the manufacture of large forging ingots. The Leybold-Heraeus Roechling large ingot development program is continuing.

C. Conclusions and Recommendations

1. Conclusions

a. An ESR manufacturing system is needed to produce small-diameter ingots (7 inches or less in diameter) at a volume, cost, and quality level comparable to existing commercial practice. The feasibility of such a system has been demonstrated.

b. Fundamental technical differences exist between fixed full-length or short collar molds that have not been evaluated fully and would have a bearing on the future design of ESR furnaces. There are certain operational advantages with each approach.

c. The metallurgical effects of electrode change events are in dispute and should be clarified in the interest of optimizing future ESR furnace designs. The electrode change technique (successive remelting of more than one electrode to form a larger ingot) commonly is used with collar mold ESR furnace systems and, to a lesser degree, with furnaces using fixed full-length molds.

d. There currently is no ESR hollow ingot production in the United States although capability is being developed. The Soviets have experimental ESR furnaces producing moderate-diameter (up to 20 inches) hollow ingots. The Japanese also have a facility for making thin-walled small-diameter cast tubing.

e. Ingot mold design and performance is a continuing problem with existing commercial ESR facilities. The problem, while moderate for round cross-section molds having diameters of approximately 50 inches or less, intensifies as the mold diameter increases and is quite serious with high-aspect-ratio rectangular molds used to cast large slab ingots. Unless brought into proper perspective, slab mold costs (capital and maintenance) will limit the size of slab ingots that could be made by the ESR process.

f. Substantial heat is lost to the water-cooled system at the slag/mold wall interface in ESR processing. No method has been developed to reduce this loss; however, significant electrical power savings could be realized if such a system could be developed.

g. The design and operation of adequate emission control apparatus are not problems because of the relatively low volume of emissions from ESR furnaces. Environmental factors are considered thoroughly prior to the design and installation of each ESR facility.

h. The absolute power required to operate all existing and proposed ESR furnaces is so small relative to the total energy requirements of the steelmaking industry that its energy burden (less than 0.3 percent) is insignificant.

i. The use of liquid metal feed instead of solid cast or wrought electrodes in ESR processing has significant appeal for certain applications. No commercial installations are known to exist although some experimental efforts are being conducted.

j. Hydrogen poses a major problem and the development of a reliable control mechanism must continue to be given high priority if large ESR ingots are to be successfully produced.

2. Recommendations

a. Develop an economical process for the manufacture of small-diameter ESR ingots with quality levels at least equivalent to those possible with existing large ingot commercial practice.

b. Compare and evaluate the surface and interior structure of production-scale ingots of various sizes and alloy types cast in fixed and collar molds.

c. Evaluate the influence of electrode change events on ingot solidification behavior, particularly with segregation-sensitive alloys on production-size ESR ingots.

d. Continue hollow ingot development to the point at which commercial feasibility is demonstrated.

e. Develop round (over 50 inches in diameter) ESR molds and large, rectangular-section, high-aspect-ratio slab molds of substantially improved design.

f. Develop a technique to conserve the heat lost at the slag/mold wall interface of an ESR mold, perhaps in an ingot-withdrawal collar-type mold or perhaps by employing a feature that allows a heavy insulating frozen slag layer to exist between the molten slag zone and the cooling medium.

g. Explore the feasibility of using liquid metal feed as an alternative to conventional solid electrodes in ESR processing possibly employing plasma to maintain the temperature of the slag layer.

VIII. PLASMA ARC MELTING TECHNOLOGY

A. Introduction

The transferred arc plasma jet heating recently found useful applications in the basic metallurgical industry for melting metals, superheating and refining molten metal, and drip melting consumable electrodes in conjunction with progressive solidification of ingots and shape castings.

A plasma jet is an arc-gas device capable of generating extremely high temperatures. No combustion is involved. The plasma is the ionized gas created in an electric arc discharge in which electrons, positive and negative ions, and atoms are found. The discharge is characterized by intense luminosity, since it is a good conductor of electricity and is affected by magnetic fields.

Early development and the application of plasma arc heat for melting metals and alloys apparently occurred at the United States facilities of Union Carbide Corporation. The development of certain specific and useful transferred-arc plasma torches (Gage, 1957) and high-voltage arc plasma generators using nonconsumable hollow copper electrodes (Baird, 1965) led to various concepts of plasma arc melting. Successful plasma arc melting in pilot scale by Linde Company, a division of Union Carbide Corporation, triggered intensive developmental work in this area in the German Democratic Republic, Japan, the Soviet Union, and Belgium (McCullough, 1962).

B. Plasma Arc Melting

1. Plasma Heating Devices and the Mechanism of Heat Generation

There are two basic types of plasma devices; one uses an arc between two electrodes to generate the plasma while the other does not employ any electrodes and electrical power is coupled through an induction coil. A detailed discussion of various plasma devices based on these two principles may be found in the literature (NASA, 1966).

For melting purposes, the transferred-arc-type plasma torch usually is preferred. Plasma torches of older design contain a stick electrode (Figure 35) while more advanced designs contain a hollow water-cooled copper cathode (electrode in case AC power is used). The latter type is preferred because of its higher power input capability and the noncontaminating characteristic of the heat delivered.

The electrical energy is changed within the plasma arc into other forms of energy – principally heat – and the heat is transferred by the mechanisms of conduction, convection, radiation, and diffusion. Conduction occurs by means of interparticle transfer from a region of higher temperature, called the source, to a region of lower temperature, called the sink. Radiation requires no medium since it travels as waves. Convection depends upon the difference in mass density of the heated gas and the main body of the surrounding gas. Diffusion depends upon the concentration of molecules, atoms, ions, electrons, and thermal gradients.

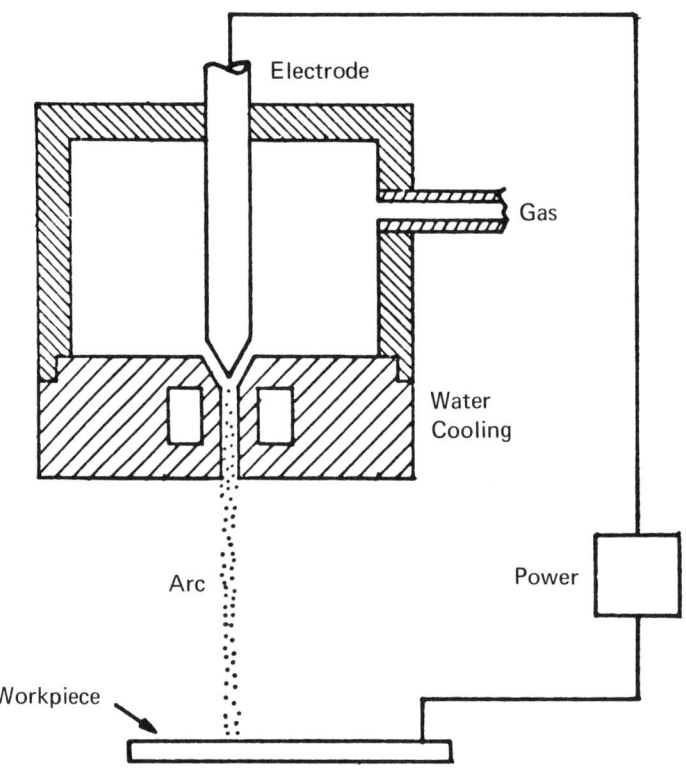

FIGURE 35 Direct Current Transferred-Arc Plasma Torch Schematic (from Bhat, 1972).

2. <u>Plasma Torch Electrode Materials</u>

Graphite, tungsten, and copper are the three basic materials used commonly as electrodes in plasma torches.

While a graphite electrode requires no cooling, it is consumed in the plasma generation process and, therefore, electrode feeding devices are required. Also, the ablated graphite electrodes are a source of plasma stream contamination that can be reduced to negligible levels in certain types of plasma furnace.

Tungsten is more commonly used than graphite as a cathode but rarely as an anode in plasma devices. Tungsten is compatible with all inert and reducing gases but suffers severe corrosion and erosion in an oxidizing atmosphere. The excellent thermionic emission and high melting point of tungsten make it an ideal cathode material. Copper is the anode material commonly used with a tungsten cathode.

Copper can be used as the anode and cathode in plasma devices. The use of copper as a cathode involves certain risks requiring a highly efficient cooling system for the protection of the hollow cathode and the nozzle.

3. Choice of Gas

The choice of plasma gas is important. The noble gases like argon and helium have low ionization energy and reasonable arc voltage. Properties of interest in selecting a gas are listed in Table 29. If long anode and cathode service life is desired, inert gas must be used to prevent oxidation of the electrodes. If a low-voltage power supply is desired, a monatomic gas must be used in the plasma torch. In industrial installations where both operating cost and long service life of electrodes are necessary, the choice is an inexpensive inert gas.

TABLE 29 Properties of Gases Used in Plasma Applications

Gas	Dissociation Energy (Kcal/g Mole)	Particle after Dissociation	Ionization Voltage (Volts)	Open Arc Voltage[1]	Arc Temp.[2] (°F)	Heat Content of Gas[2] (Watts/Ft^3 [NTP])
A	0	A	15.68	18	18,000	75
He	0	He	24.46	26	27,000	110
H_2	104	H	13.53	70	15,000	260
O_2	110	O	13.55	40	16,000	425
N_2	225	N	14.48	40	16,000	425
Air	--	--	--	60	16,000	425

[1] Constant Arc Length and Power. [2] 10 percent thermal ionization.

SOURCE: Bhat, 1972.

Figure 36 plots the plasma temperature as a function of gas energy content for several plasma gases. Note that dissociation and/or ionization of the gases has a direct influence on the temperatures developed. The composition curves of the gas plasma indicate that nitrogen is an excellent heating medium because the dissociation state change for nitrogen occurs at lower temperatures than diatomic gases. However, argon plasma transfers heat efficiently, and nitrogen and argon are often mixed with hydrogen when used as plasma heating gases. Hydrogen raises the voltage of the arc to permit the use of higher power without an appreciable increase in the current levels. The hydrogen addition increases the thermal efficiency and also the thermal conductivity of the plasma arc. The presence of hydrogen also reduces the chance of oxidation of the heated body or the metal during melting.

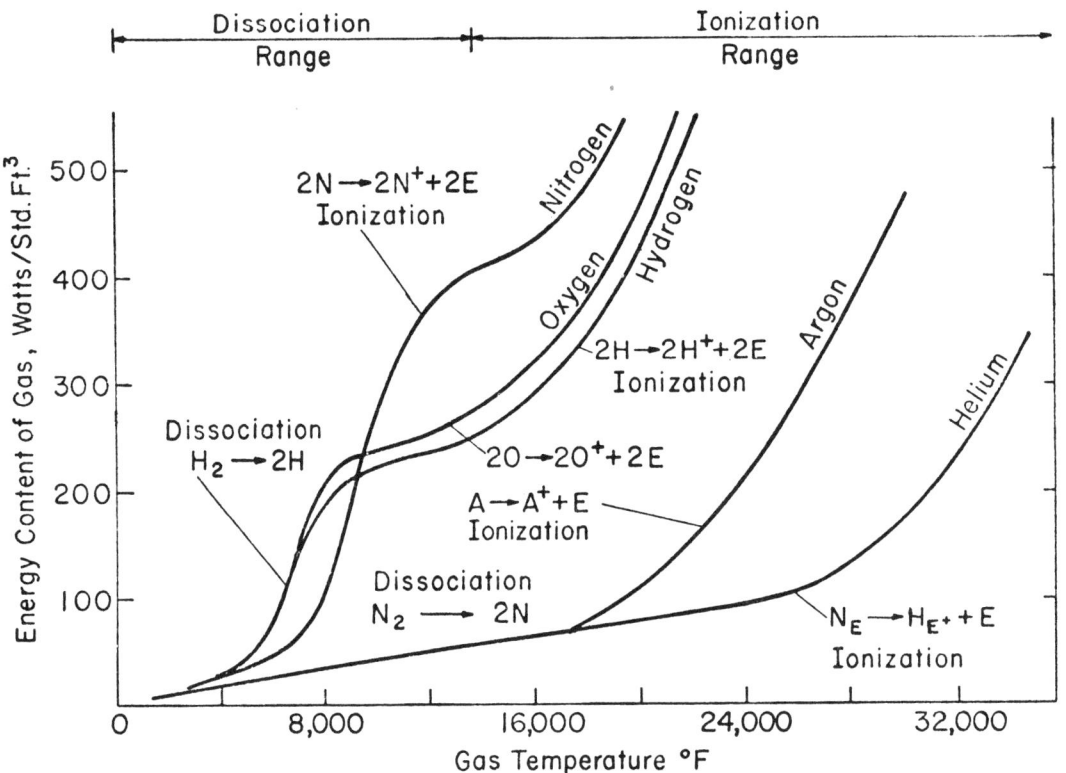

FIGURE 36 Plasma Temperature as a Function of Gas Energy Content at Atmospheric Pressure (from National Aeronautics and Space Administration, 1965).

4. Application of Plasma Heat for Melting and Remelting of Metals and Alloys

Plasma heat sources are characterized by high concentration of energy and flexible control of heat is possible. Also, plasma arc remelting is characterized by highly stable electric power conditions. Plasma heat provides an independent adjustment of electrode melt rate and heating of the liquid metal bath.

The objectives and incentives for using plasma heat for melting, refining, and ingot casting vary for the different adaptations of the process technology (melting systems) and the alloys to be processed. Plasma melting processes may compete with or complement primary melting processes such as electric arc furnace melting, vacuum induction melting, and secondary remelting processes (such as nonconsumable electrode melting and remelting, electroslag remelting, vacuum arc remelting, and electron beam melting). Plasma melting process adaptations simply offer different choices.

These choices enhance the capabilities of the plasma arc melting process to provide for:

1) Minimal or no contamination depending upon the containment receptacle used to hold the molten metal
2) Compositional control and consistency from heat to heat
3) Rapid refining ability using minimum amounts of refining agents
4) Ability to process different types of metals and alloys
5) Control over ingot solidification
6) Productivity, process simplicity, reliability and economy

Much of plasma melting development on an industrial scale reportedly occurred in the Soviet Union, Japan, and the German Democratic Republic. (Reports on these developments are sketchy; however, salient features of these plasma arc melting and ingot solidification systems are described in section C.) Except for one project directed towards implementation of plasma arc melting as a production process for the manufacture of improved structural materials, there has been limited activity in industrial scale plasma melting in the United States (U.S. Air Force, 1971).

C. Plasma Arc Melting Furnaces

Since plasma arc melting of metals and alloys is accomplished in several ways, the plasma melting furnaces can be set up for primary melting as well as remelting. The charge material for primary melting may be virgin metals, alloy scrap, or a mixture of these materials. Primary melting plasma furnaces have a refractory lining similar to that in the electric arc or the vacuum induction furnace.

A plasma induction furnace is one in which the plasma arc heat is used as auxiliary heat for speeding up the melting of charge materials by electric induction in a refractory crucible.

Another type of plasma arc furnace utilizes a short water-cooled copper crucible in which graded alloy scrap is charged and consolidated into an electrode for subsequent remelting by any of the consumable electrode remelting processes (electroslag remelting, vacuum arc remelting, electron beam remelting, or plasma arc remelting).

Plasma arc heat also is used for remelting of electrodes by drip melting and ingot casting through a cold-wall sleeve mold by the withdrawal technique. Other plasma melting systems include cold hearth melting, skull melting, and shape casting furnaces. As yet, these plasma furnaces are not viable for industrial applications.

1. Linde Plasma Arc Furnaces

Early plasma arc melting furnaces employed transferred arc plasma jets with stick electrodes (McCullough, 1962). In later furnaces, plasma jets used hollow copper electrodes (Baird, 1965). The plasma arc furnace, shown in Figure 37, is similar in shape to and may employ the same refractory lining as the conventional electric arc furnace with graphite electrodes. The arc from the plasma torch is transferred to a water-cooled copper electrode fitted in the bottom of the hearth. To prevent contamination of the argon atmosphere over the melt, the roof-side wall joint of the furnace is sealed with a labyrinth sand seal through which the effluent gas can escape. The molten metal stirring was accomplished by using an induction stirring coil. The first two pilot furnaces of this type melted 300 pound and 2,000 pound heats.

Experimental melting in these early plasma arc furnaces used both relatively pure raw materials and mill scrap to manufacture various alloy steels, tool steels, bearing steels, and superalloys. The possibility of manufacturing materials having gas contents and physical properties approaching those produced by vacuum melting was indicated.

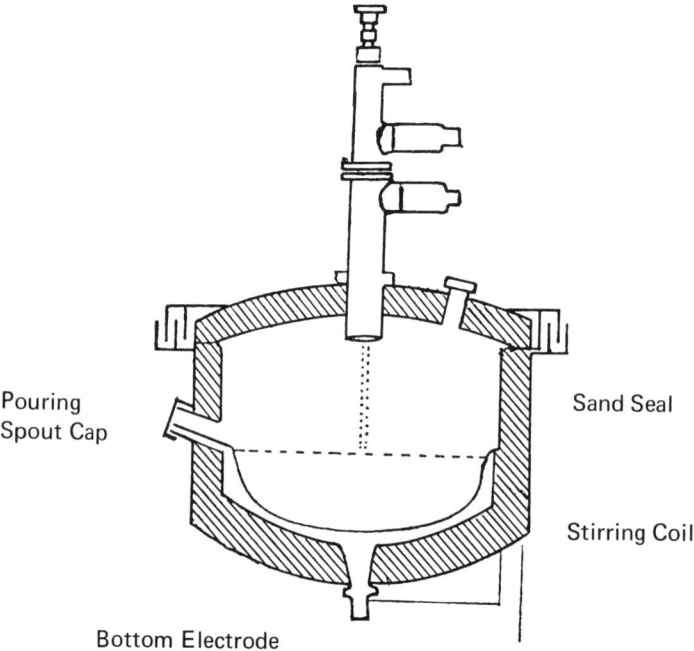

FIGURE 37 Cross Section of the Plasma Arc Furnace (from McCullough, 1962).

2. Scrap Recycling Plasma Arc Melting Furnace

In the German Democratic Republic, a plasma melting furnace was designed, constructed, and operated to process 10-ton charges of nickel alloy scrap, stainless steel scrap, superalloy scrap, and tool steel scrap (Lachner et al., 1973; Fiedler et al., 1974). Plasma melting of scrap is performed in a furnace whose walls and roof are lined with rammed chrome-magnesite bricks. The furnace is fitted with two plasma torches in the side wall and one in the roof. Each torch is fed with 6000 amperes at 100 to 600 volts. Reportedly, the current-carrying capability of the plasma torch at this facility recently was raised to 9000 amperes DC. Argon and argon-nitrogen mixtures are used in the plasma torch and the argon consumption per torch is around 28 cubic ft/hr.

The service life of the plasma furnace lining is 60 to 80 heats. The furnace roof heat is controlled through a radiation pyrometer mounted on the side wall and a thermocouple probe in the roof. When the side wall and roof temperature reach 3362° F the power input is reduced automatically, thus increasing the service life of the furnace side wall and the roof. This plasma furnace reportedly has the capability to melt a scrap charge at the rate of 6 to 9 tons/hr. The specific power consumption claimed for the melt-down period ranges from 600 to 800 kWhr/ton. The specific power consumption reported for the finish-melt period of the alloy is an additional 100 to 200 kWhr/ton.

The reported alloy yields are as follows:

Metal	Percent
Nickel	100
Molybdenum	100
Tungsten	100
Chromium	97-100
Manganese	96-100
Silicon	70-85
Aluminum	50-70
Titanium	20-40

The hydrogen content of plasma heats made in this furnace is said to be under 0.30 cubic inch/3.5 ounces (avdp.) (5 cm^3/100 g) with typical hydrogen content around 0.12 cubic inch/3.5 ounces (2 cm^3/100 g). Oxygen content is noted to be less than 20 ppm and the nitrogen content of these steels is said to be the same as that in electric arc furnace melted grades.

Plasma arc melting in refractory-lined vessels of virgin charges, all scrap charges, and mixed charges of virgin materials, oxides, and scrap is claimed to produce an alloy quality comparable to that produced either in vacuum induction or vacuum arc remelting furnaces. Limitations in achieving high quality are said to be due to casting ingots in conventional molds in air.

Plasma arc melting in refractory-lined furnaces may be used in the industry to:

1) Melt scrap of nickel-base alloys, iron-nickel-base alloys, iron-nickel-chromium alloys, and superalloys
2) Remelt scrap of high-speed steels, bearing steels, and nitrogen-alloyed steels
3) Produce magnetic and resistance alloys.

3. Plasma Induction Furnace

The Daido plasma induction furnace (Figure 38) uses a water-cooled roof, magnesia-lined crucible encased in a steel tank (graphite lining used for melting copper alloys) (Asada and Adachi, 1971). The plasma torch is the arc-transfer type with a maximum current said to be 2300 amperes DC. A recent new torch design raised the maximum current to 5000 amperes. The amount of argon gas varies with the stability of the plasma arc and economic considerations and is 17.5 to 21.0 cu. ft/hr (5 to 6 m^3/hr) under normal operating conditions of around 2000 amperes.

The electrode is of thoriated tungsten and the water-cooled anode is made of copper. The electrode at the bottom of the induction furnace is graphite; however, it is not in direct contact with the molten metal and carbon contamination is avoided.

The plasma induction melting and the resulting material evaluations performed and reported by Daido Steel (Table 30) indicated that:

1) Plasma induction melted material generally is equivalent to that produced in an industrial-size vacuum induction melting furnace
2) Plasma induction melted material is superior to metal that is produced in an electric arc furnace and subsequently vacuum degassed
3) Plasma induction melted material is suitable for the manufacture of castings of heat-resisting alloys and electrodes for remelting
4) Alloy element recovery in the plasma induction melted metal is greater than 95 percent (Table 30)
5) Inclusions found in plasma induction melted heat-resisting alloys containing large amounts of titanium and aluminum generally are smaller than 10 microns (10μ) in size.

TABLE 30 Alloy Recovery in Plasma Induction Melted Metal (Daido Steel Data).

Element	Yield of Alloy Addition (%)
C	100
Si	99
Mn	98
Cr	100
Al	96.5
Ti	95.5
V	100
B	95.0
Nb	97.0

SOURCE: Asada and Adachi, 1971.

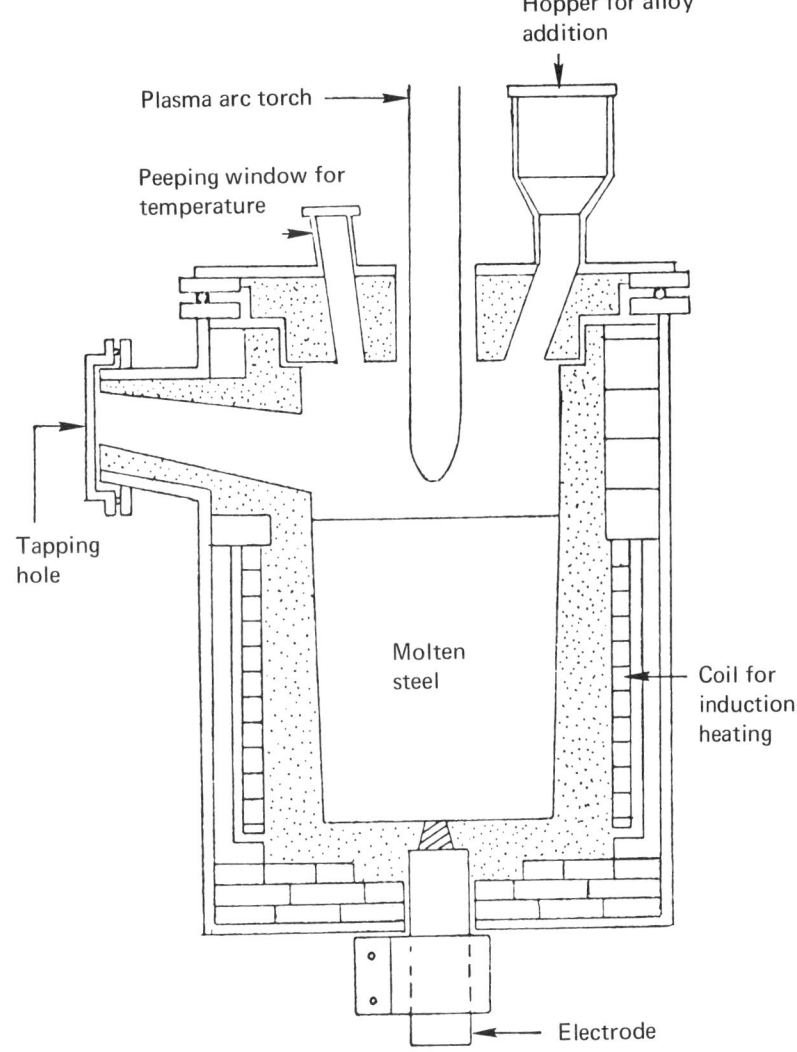

FIGURE 38 Plasma Induction Furnace (from Asada and Adachi, 1971).

4. <u>Plasma Arc Remelting (PAR)</u>

In the plasma arc remelting process that is used at the Paton Institute, Kiev, USSR (Paton et al., 1971), consumable electrodes are remelted using two or more plasma heat sources (plasmatrons). The essential features of the Soviet plasma arc remelting process are shown in Figure 39. The ingot is solidified in a water-cooled copper mold that is fitted with a continuously withdrawn bottom plate. The electrode is fed down with a rotating motion. The

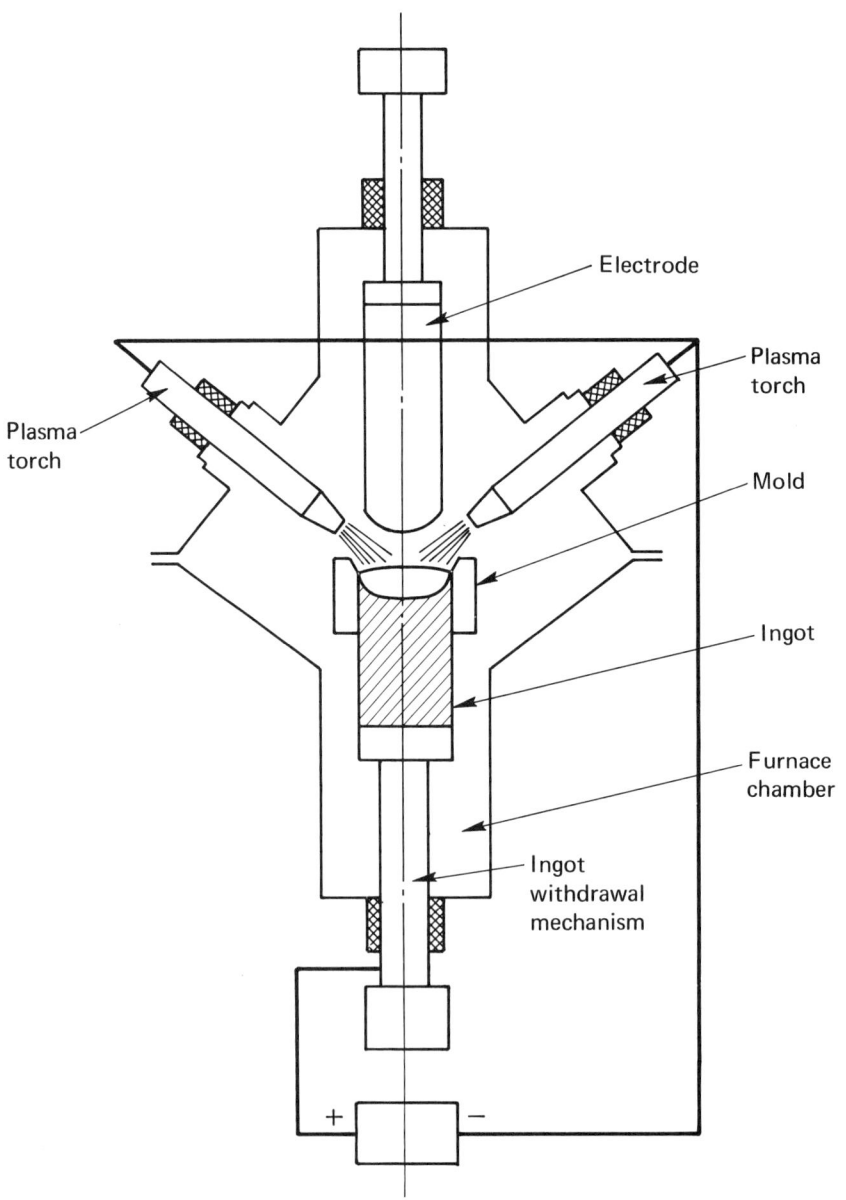

FIGURE 39 Technological Scheme of Plasma Arc Remelting – Soviet Design (from Paton et al., 1971).

plasma jets impinge on the electrode end, and metal transfer to the pool occurs in the droplet form. The use of several plasmatrons is said to offer close and uniform control of the molten metal superheat. As a result, the Soviets claim that a flat and shallow metal pool is obtained regardless of the heat conductivity of the metal being remelted.

Heat transferred to the charge by the plasma guns of Soviet design (Paton Institute) is claimed to be in the range 65 to 85 percent. These plasma guns are stated to operate consistently in a wide pressure range from 0.39^{-2} inches (10^{-2} mm) of mercury up to several atmospheres. The loss of elements through evaporation thus can be controlled to within narrow ranges.

Certain other possibilities of Soviet plasma arc remelting may be summarized as follows (Lakomsky, 1972; Savitsky et al., 1974):

1) Iron-nickel alloys can be deoxidized using hydrogen, resulting in ingot metal containing no deoxidation products
2) Ultra-clean bearing steels for the manufacture of miniature bearings can be produced using slags of different types to adjust the quantity and types of inclusion obtained in the resulting ingot
3) Up to 1.0 percent by weight nitrogen can be added to certain steels through pressure remelting
4) Ingots with dendrite orientation nearly parallel to the ingot axis – a feature claimed to be very useful in hard-to-work materials, titanium metal, and titanium alloy ingots for direct use in the manufacture of tubular shapes – can be produced
5) Titanium and titanium alloy ingots can be produced directly from lump sponge and sponge-alloy mixtures
6) Fairly large (2 inches diameter) monocrystals of high-melting-point metals (e.g., tungsten, rhenium, molybdenum) can be produced.

Soviet reports also indicate that noble metal alloys, when plasma remelted, are refined as in electron beam furnaces and that the loss of metal through evaporation is negligible during plasma arc remelting. Soviet research in plasma arc melting process appears intensive in the area of gas exchange processes at the interfaces between the gas atmosphere-liquid metal and liquid metal-solidified metal.

Plasma arc melting in the Soviet Union, Japan, Belgium, and the German Democratic Republic appears to be a well funded, independent metal processing technology rather than a strategic or competitively evolved process technology as it is in the Western Hemisphere. Therefore, few of the Soviet claims in reports of plasma arc remelting studies were verified through similar studies conducted elsewhere.

In the United States, the impetus and technical justification for development of plasma melting processes involved the need to recycle nickel-base alloy, titanium alloy, and other reactive metal alloy scrap with no loss of alloy ingredients and no degradation of quality or intrinsic market value. This need prompted development of a novel dual-duty plasma arc furnace (U.S. Air Force, 1971). A major portion of the development work was devoted to production of a nonconsumable rotating electrode melting system (Lowry and Schlienger, 1971).

The plasma arc remelting installation at Schlienger, Inc., San Rafael, California, can produce ingots 14 inches in diameter using a single 1 MW torch connected to a 600 kW output DC power source. The ingot is withdrawn from a short water-cooled copper sleeve mold as shown in Figure 40. The use of a water-cooled copper electrode instead of a tungsten or other electrode in the plasma torch avoids melt contamination by the ablated electrode material. The water-cooled copper mold eliminates refractory contamination of the melt.

This furnace also can prepare electrodes through scrap consolidation using plasma heat and continuous withdrawal of the fused alloy aggregate in the water-cooled mold. An attractive feature of this dual-function plasma melting system is its ability to recycle a 100 percent scrap charge of alloys containing large amounts of reactive elements (e.g., aluminum, titanium, and silicon) and easily oxidizable elements (e.g., manganese, vanadium, and niobium).

Plasma arc remelted ingots of high-alloy steels, superalloys, and titanium alloys produced in the Schlienger plasma arc remelting equipment have shown predominantly parallel orientation of grains to the ingot axis. The ingots produced in this furnace have been relatively short (15 to 38 inches in length). Nonetheless, these observations regarding the structure of the plasma arc remelted ingots support similar findings reported by Soviet technologists. Material characterization studies are required to provide further explanation of the various structural features noted in plasma arc remelted ingots.

5. Plasma Beam Furnace

The plasma beam furnace developed by Ulvac Corporation, Japan, is a hollow cathode discharge melting system (Bhat, 1972). The plasma beam heating principle is shown in Figure 41. A hollow cathode and a workpiece to be heated (anode) are located opposite each other in a vacuum chamber. The chamber is evacuated down to 1×10^{-3} torr. The cathode and anode are connected to a DC power source with a high-frequency starter power source superimposed on the main electrical circuit.

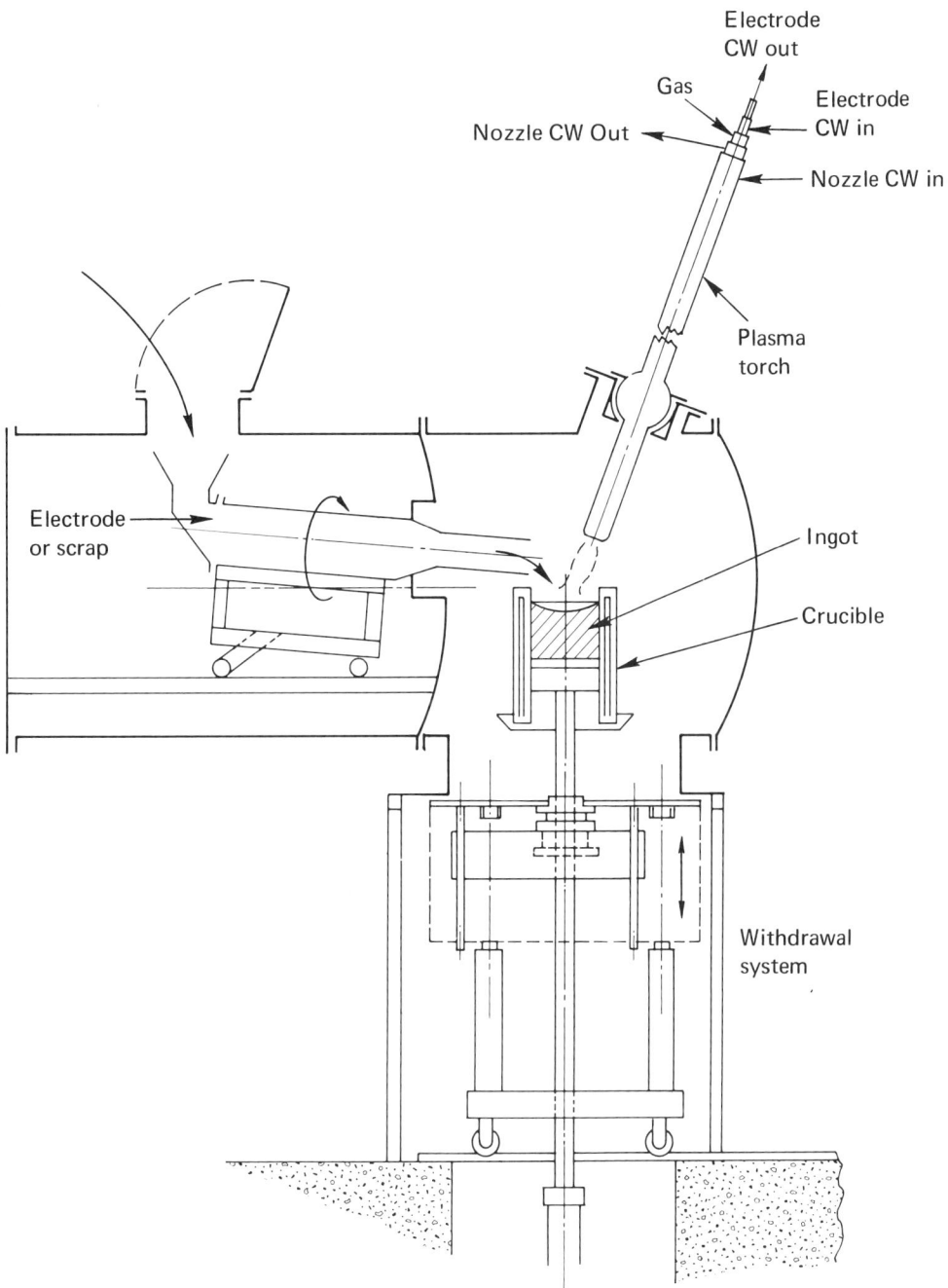

FIGURE 40 Schematic of Schlienger Plasma Melting-Remelting System (from Air Force Materials Laboratory, 1971).

CW = cooling water.

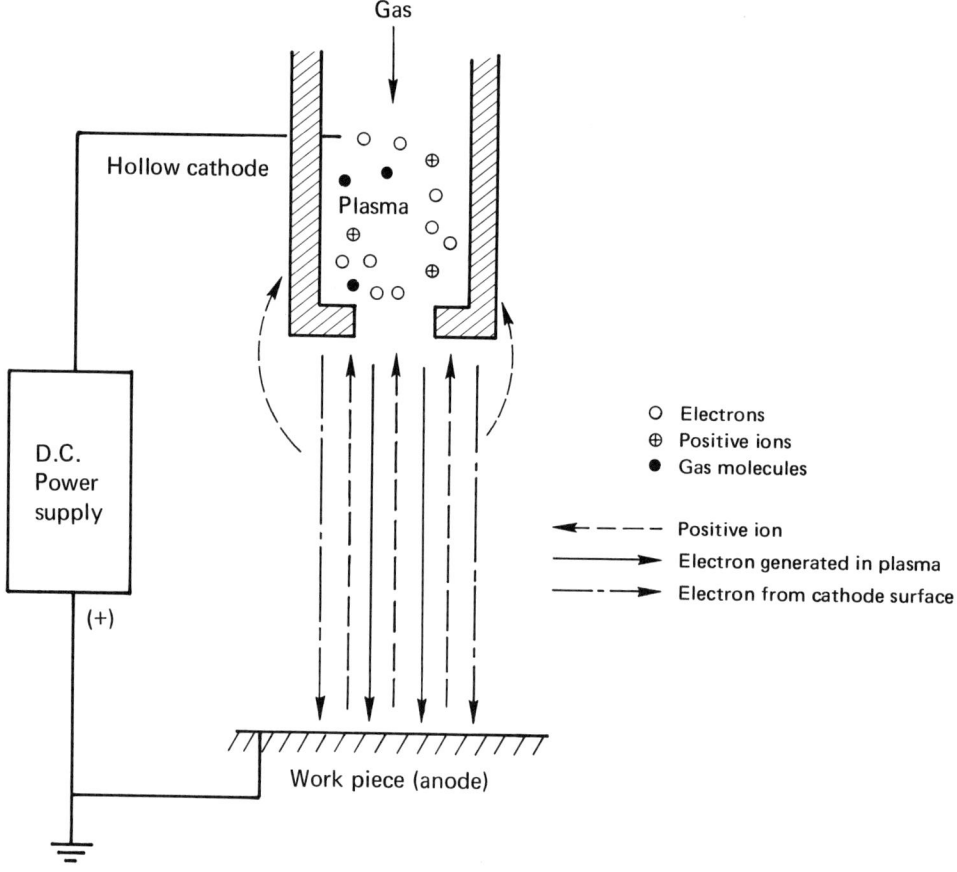

FIGURE 41 Schematic of the Heating Principle of the Plasma Electron Beam (from Bhat, 1972).

Simultaneously with the closing of the electrical circuit, a small amount of gas (usually argon) is introduced into the cathode. In the presence of the high-frequency electrical arc, the gas molecules in the hollow cathode are ionized to generate a low-pressure gas plasma. Positive ions in low-pressure gas plasma raise the temperature of the cathode up to 3860° F by bombarding the internal diameter surface of the cathode. When the temperature of the cathode is sufficiently high, thermionic emission occurs and the electron density around the cathode increases rapidly. When the main power is superimposed, plasma in the cathode and emitted electrons from the hollow cathode are accelerated to the anode that produces an intensely hot beam for either heating or melting the workpiece.

The electrical characteristics of the plasma beam are affected by:
1) Type and amount of gas introduced into the hollow cathode
2) Distance between the electrodes
3) Pressure of gas in the chamber
4) Shape and dimensions of the cathode

Plasma electron beam (E/B) has the following advantages and disadvantages when used as a heating source. The advantages are that:
1) It is a highly stable heating source
2) Thermal efficiency ranges from 65 to 90 percend depending upon the distance between the electrodes
3) There is no danger of X-ray generation since the voltage used is in the range of 20 to 100 volts
4) Furnace costs are lower than those for conventional E/B furnaces
5) Plasma beam shape is adjustable
6) No contamination of the melted or heated metal occurs
7) Alloys containing reactive elements can be melted because a lower vacuum is used in plasma E/B melting.

Disadvantages of plasma E/B heating are that:
1) High energy cannot be concentrated on a small spot because of the use of low voltages and high currents
2) Very pure inert gas is needed since impurities in the gas contaminate the heated (melted) metal.

Plasma E/B melting furnaces reportedly are suitable for the melting of reactive metals such as titanium, zirconium, niobium, molybdenum, tantalum, and tungsten. A one-step titanium slab ingot manufacturing plasma beam furnace was designed and operated by Ulvac Corporation for Nippon Stainless Steel Corporation and underwent preliminary melting trials. Both slab ingots and 18-inch-diameter round titanium ingots were produced in one step from sponge to ingot (Bhat, 1972).

The projected cost of plasma beam melting large titanium slab ingots is approximately $200/ton. This plasma beam slab ingot production furnace reportedly is undergoing pre-production trials.

D. Plasma Melting Process Economics

The economics of various plasma melting processes are yet to be determined through large-scale industrial application studies. Preliminary reports of operating small-size (1/2 to 2 tons) plasma induction furnaces indicate that plasma induction melting costs are slightly higher than electric arc furnace costs but lower than those of similar size vacuum induction furnaces. The argon gas is recycled in Daido plasma-induction furnace. Soviet reports indicate that plasma arc remelting costs are not higher than vacuum arc remelting costs for installations of the same size.

E. Conclusions and Recommendations

1. Conclusions

    a. Plasma melting technology is still in the early developmental stage.

    b. Equipment builders are evolving suitable designs for the application of plasma heat for primary melting, molten metal refining, and consumable electrode remelting purposes.

    c. Data on the following are insufficient:

        1) Design requirements of plasma torches for melting operations

        2) Plasma torch operating parameters

        3) Plasma torch operating power sources

        4) Plasma gas recirculation

        5) Plasma arc furnace designs and specifications

        6) Plasma arc melted and remelted material characterization.

    d. No economic appraisal of industrial-scale plasma melting and plasma remelting installations using a variety of experimental charge materials has been done as yet.

2. Recommendations

    a. Technology relative to transferred arc plasma torch design (construction and operation) for operations such as an auxiliary heating source in melting operation and ladle refining should be developed further.

    b. Industrial-scale plasma melting projects should be established to study the potentials of gas-metal refining and slag-metal refining under the high-temperature environments created by plasma arc heat.

c. Industrial-scale plasma arc remelting studies using various input materials should be conducted to permit full evaluation of the benefits that could be derived from differences in the process features (e.g., heat transfer, ingot solidification, and gas-metal reactions).

d. The possibilities for application of plasma arc heat in the production of large monocrystals of various refractory metals (e.g., tungsten, molybdenum, chromium, ductile alloys of chromium) and the melting of ceramic materials (e.g., oxides, carbides, and nitrides of various metals) should be identified.

IX. CRITICAL EVALUATION OF ELECTROSLAG AND PLASMA ARC MELTING

A. Electroslag Melting

1. General Evaluation

The range of actual and potential applications of electroslag melting is very wide, but each application must be carefully evaluated. As has been indicated throughout this report, ESR generally provides some improvements in virtually every area in which it has been used. The extent varies, and there are the anticipated exceptions. The notable successes have been widespread in alloy grades and sizes. For example, the superalloy field has been successful in improving the workability of several traditionally difficult alloys, notably gamma prime (λ') strengthened alloys. Also, a distinct improvement was seen in mechanical property reproducibility in most grades. On the other hand, the early expectations of substantial absolute improvement in properties over VAR were not fulfilled. Indeed, in the related area of high-speed steels, the improvement in properties over good air-melt practice is barely discernible. The pitfalls of expecting the process to produce property benefits in all uses are well illustrated by that example. ESR substantially improves a poor-quality air-melt material but will not give the same incremental improvement over materials that are already of high caliber before remelting.

The ESR method provides a useful route to large high-quality ingots primarily for the energy-related industries. Its industrial value is indicated by the significant number of installations being commissioned. When viewed in the correct context, the products of large ESR furnaces are of high quality.

This report endeavors to collate the available information in a dispassionate manner and it must be read with that in mind. The confusing claims and exaggerations must be sifted from the solid evidence of progress and potential in the ESR method. Where possible, the dubious claims have been extracted and set against the available evidence. Also, the generally-accepted advantages were presented in the context of the appropriate end-use and alloy grade.

2. High-Alloy Materials

Possibly the most exaggerated claims for ESR benefits were made in the early stages of process development. Some of the general statements were shown subsequently to be qualitatively true but others remain highly suspect. Dealing first with the suspect category, let us first consider the claimed substantial improvement in carbide distribution in high-speed steels. Upon investigation it appears that this claim is unsubstantiated, and only marginal structure differences are observed between ESR, VAR, and good air-melt high-speed steels. The observed property improvements are due primarily to differences in inclusion content and improvements in the structural uniformity of the ingots. The claim also has been made that conventionally unworkable superalloy compositions may be forged after ESR melting. Some differences in hot-working behavior have been reported between ESR and VAR but the difference is primarily an effect observed in hard-to-work alloys.

The economic aspect of the higher melt-rates acceptable in ESR as opposed to VAR has been overstated frequently since the present lower efficiency of ESR together with slag costs counterbalance the advantage. The advantage of line-frequency power also has been overstated since the need for precise control and the problem of supply balance outweigh the disadvantages posed by a modern frequency-converter system.

The ESR furnace was claimed to be a simpler machine than its VAR counterpart. However, many of the early, simple machines were disastrous failures for their purchasers since they possessed insufficient control and mechanical reliability for efficient working. A modern ESR furnace is at least as complex as its VAR equivalent. Crucible problems supposedly were solved by VAR technology prior to ESR design; however, the ESR problems are much more severe, particularly in shapes other than round, and most ESR operators regard these problems as far from solved. The economics of making long ingots were said to be greatly improved in ESR over VAR due to the possibility of electrode changing; yet, years after the concept was put into operation, controversy is still extensive over the possibility of defects introduced by this operation.

The alloy element control problem, particularly in λ' nickel-base alloys, has been "solved" repetitively by the use of complex slags but still is a troublesome problem. Also, very little information exists on the behavior of residual trace elements during ESR processing.

Inevitably, a new process in a highly competitive area of proprietary alloy-making is subjected to conflicting claims. Fortunately, areas of general agreement can be extracted from the available evidence. The first of these is improved surface quality. This property is invariably higher in ESR than in other processes except possibly Durville casting. The better surface is produced by the slag skin on the crucible wall together with the extremely uniform advance of the slag/metal interface relative to the ingot base. There is disagreement, however, concerning the benefit of improved surface quality. The direct metallurgical benefit is probably small except in some extremely brittle alloys. The indirect effect on yield and productivity varies widely from operator to operator but is always beneficial.

The ESR process is less prone to solidification defects than is VAR. The more stable melting, larger thermal inertia, and probable smaller electromagnetic effects all contribute to this advantage. ESR certainly does not eliminate defects in a segregation-sensitive alloy but it does allow safe production at substantially higher melting rates. The ESR process does not produce lower oxide inclusion contents than VAR but can achieve contents that are within VAR specifications. ESR will not remove hydrogen and nitrogen but can remove effectively much larger amounts of sulfur than VAR does. Probably, with the requisite electrode quality in both cases, a VAR and ESR ingot may be made with the same general chemical composition and structural features in many alloy grades. However, in normal practice, the key differences lie in:

1) Ingot shape
2) Surface quality
3) Ingot structure
4) Production rate
5) Electrode cost
6) Melting rate.

The electrode requirements for dimensional accuracy, surface quality, and metallurgical conditions are less severe for ESR than for VAR although, in high-quality products, the restrictions on electrode deoxidation are equally close in both processes. In general, ESR furnaces permit simpler handling and turn-around than do VAR furnaces, although USSR designs of ESR equipment are an exception to this rule. The melt rate is higher in ESR and so overall productivity is substantially greater. If this productivity is used, then ESR is a cheaper and technically adequate route. In a few cases, there are reports of variants in ESR processing that allowed the manufacture of ingots that otherwise could not have been manufactured without great difficulty. In this category are materials with precise, small additions of a minor component (e.g., a rare earth), and this aspect has significant development potential, particularly with respect to ingot inclusion composition control.

The ESR and VAR processes utilize cold crucibles to provide advantageous macrostructures and, therefore, are inefficient from an energy standpoint. Many years of VAR development failed to improve its efficiency significantly even though such improvement may not have been sought specifically. Schemes of slag-composition modification, crucible modification, etc., may lead to greater inefficiencies which would be desirable from the viewpoint of ESR process economy but the total energy used by all remelting processes is extremely small.

In summary, the potential of ESR over VAR in the high-alloy fields is mainly in economic improvements from: slabs and hollows as opposed to rounds, in-house yield and productivity gains, and slightly better return on investment. The potential technical advantages lie in the use of ESR to effect unique composition control on the ingot, specifically with respect to inclusions. In the latter case, the matching of a deliberate inclusion composition to the required end-product properties has a very considerable potential, obtainable only with ESR or plasma/slag processing.

3. Low-Alloy Materials

The assessment of ESR for low-alloy materials has been slow because of the more critical economic evaluation applied. The needed ingot size in these alloys is generally much larger than in the high-alloy field and the increment of cost allowed for mechanical property improvement is very small. The advantage of ESR over VAR becomes significantly clearer in this case since the difference in production rate is more pronounced in the alloys less sensitive to segregation and in the larger ingot sizes. The inherent stability of the ESR method on large electrodes is also very important.

The largest VAR furnace presently in production (5 feet ϕ) probably represents the maximum stable process size even when melting at half the rate of the equivalent ESR furnace. For large ingots, then, the ESR process and its variants probably represent the only practical way of obtaining the necessary improvements in ingot structure by remelting. In this context, ESR is not to be regarded as an alternative to VAR since the latter method is impractical in these sizes and has too low a potential production rate for economic viability. The question surrounding ESR is whether the method itself is warranted, or economical, in low-alloy materials.

Over the past few years, techniques were developed for manufacturing low-sulfur steels economically in conventional equipment (e.g., ladle refining methods). In view of these developments, one of the reasons for using ESR — desulfurization — loses some of its incentive. Any change in slag composition that reduces the melting rate or slag resistivity to attain the high possible desulfurization is costly. Hence, two questions must be answered.

1) Is ESR justified on ingot structure improvement alone?
2) Are additional process steps in liquid steel a cheaper method of obtaining optimum analysis than is ESR?

In order to resolve these questions, a systems approach is clearly necessary and, undoubtedly, will expose areas where ESR is the best route and areas where it is clearly uneconomical. No such examination has appeared yet in the literature.

Major problems in the ESR manufacture of large-section steel ingots appear to be those of composition control with respect to hydrogen and aluminum. In many forging grades, low levels of aluminum are required and are extremely difficult to obtain with normal ESR slag compositions. When the low-aluminum requirement is superimposed on the low-hydrogen requirement, the possible choice of slag compositions is restricted further, even with degassing of the primary electrode melt. The best combination of closed-atmosphere furnaces, slag compositions, and primary melt treatment to attain an acceptable compromise on sulfur, aluminum, and hydrogen contents at an appropriate melting rate has not been determined.

In the case of large slab sections, the mechanical property improvement has justified the several large new ESR furnaces under construction. The application of ESR to large forgings, shafts, rolls, and valve housings also is established reasonably well. In some cases, the material is claimed to be usable as-cast, and these claims should be investigated in a well-structured U.S. program since a positive finding would alter greatly the economics and availability of large steel components.

4. Reactive and Refractory Metals

Based on past experience, the application of ESR to the reactive metals probably is very limited. Titanium, zirconium, and molybdenum were investigated extensively with the finding that ESR offers no advantage over VAR although, in some cases, the ingot quality is improved over (or comparable to) that of VAR. The ESR processing of uranium and chromium promises to be simpler than vacuum induction melting but any advantage in properties for the ESR method still must be demonstrated. However, the ESR potential for making shaped castings may prove useful in this area. Clearly, in alloys where interstitial removal is a prerequisite for adequate properties, the aspect of slag contact and absence of a vacuum are distinct disadvantages.

B. Plasma Melting

1. Remelting Processes

Potentially, the availability of a heat source that is independent of metal feed rate should be most useful when unusually slow melting is required or when a solid feed electrode is not available. In adidtion, since the practical limit on plasma power density is extremely high, an application to high-melting-point systems is expected. The areas that best fulfill these categories are:
    1) Titanium alloys
    2) Refractory metals
    3) Superalloys
    4) Some highly alloyed steels.
For some specialized purposes, notably the manufacture of thin-walled tube, the PAR method has been applied to titanium alloys in the USSR. While this work apparently has been extended to other applications of similar alloys, there is only limited potential for the method in this area.

The potential control of solidification has been used only in manufacturing tungsten single crystals and has not been applied commercially to complex alloys. At present, the principal area of advantage in chemical control is in high-nitrogen stainless steels and in low-interstitial stainless steels. If the present positive indications in small furnaces can be scaled successfully to larger furnaces, this process may experience significant growth.

The scrap-consolidation capability of the PAR furnace has been demonstrated for a range of materials, including superalloys, titanium, and zirconium alloys. This method is used to produce adequate electrode stock at a high production rate.

The case for using PAR with a slag to control or remove inclusions has not yet been investigated adequately. Also, it may be possible to ameliorate some of the problems with PAR ingot surfaces by using slag as a heat-transfer modifier.

The claims for structure modification in PAR ingots of complex alloys, based on the existence of a unique PAR ingot heat-transfer mode, represent a potentially valuable development area. To thoroughly investigate the claims made for PAR in structure modification, work is necessary on a system whose structure behavior on solidification is well understood. In spite of the obvious applicability of plasma remelting to superalloys, this work has not been carried out.

2. Plasma Melting

The results on the German Democratic Republic 10-ton plasma furnace melting both steels and nickel-base alloys with relatively simple equipment are encouraging. Similarly encouraging metallurgical results also were obtained in this country and Canada on similar but smaller furnaces. Nevertheless, the data are not sufficient to permit an evaluation of the German improvements in respect to refractory life, furnace shape, and torch design. However, in view of the increased knowledge of arc furnace behavior gained in the past 10 years, the place of similar furnaces in the production scheme should be reassessed, particularly in view of the steps taken by the Japanese in the plasma induction melting equipment.

Plasma remelting has potential advantages over an ultra-high-power (UHP) furnace in melt-down rate for steel scrap by eliminating the graphite electrodes and providing very stable melting. The chemical advantages are probably marginal over the UHP/ladle refining systems now available. While the chemically inert, stable melting zone is a major advantage, no other positive advantage has been demonstrated. Accordingly, major potential applications are in melting very-low-carbon alloys, alloys with reactive components, and nonmetallic materials.

The alternative type of plasma furnace uses a water-cooled skull hearth. Such a technique may be used for the recovery of reactive metal scrap and it represents an advantageous melting alternative to EB or arc melting of reactive metals for casting. However, the problems of heat transfer severely limit the potential size of the furnace.

A frequent claim for plasma melting is that a nitrogen-containing plasma will result in liquid-metal nitrogen contents that exceed the predicted solubility limit. In a progressive freezing operation, this supersaturation is said to be retained in the ingot, leading to alloy steels with over 1 percent by weight nitrogen in the solid state. Such material may be produced on a small scale, but the claim has not been investigated adequately on a large scale. Since very large temperature gradients exist in scaled-down PAR furnaces, the effect possibly may be one of high temperature and not intrinsically associated with plasma. If this is the case, the potential merit of PAR in producing high-nitrogen steels on a large scale will disappear.

A further significant barrier to progress in the introduction of plasma equipment is the lack of adequate data on torch behavior in large melting furnaces, whether PAR or torch/refractory-furnace combinations. Disadvantages to plasma torches are known to include inadequate torch reliability, gas cost, limitations in heat transfer, and problems with magnetic interactions in multiple torch systems. Since these difficulties are of scale, their solution is unpredictable without development work on quite large installations.

APPENDIX A

GLOSSARY

Alkali fluorides -- Fluorides having group IA elements as their cationic species

Alkaline-earth fluorides -- Fluorides having group IIA elements as their cationic species

AOD -- Argon oxygen decarburization

Aspect ratio -- The length-to-width ratio of a rectangular or plate-like shape

Banding segregation -- A segregated structure of nearly parallel bands aligned in the direction of working

Cryolite -- A sodium-aluminum fluoride of the nominal composition Na_3AlF_6

Current density -- The amount of current per unit area of electrode; usually expressed in amperes per cm^2

Fill ratio -- In this report, the ratio of electrode cross-sectional area to ingot cross-sectional area; elsewhere, the diameter ratio between electrode and ingot is used sometimes

Freckle -- Spotty, usually dark macrosegregations in various arrangements, enriched in normally segregating elements and depleted of inversely segregating solute elements

Halides -- Compounds having elements in group VIIA as their anionic species

Heavy metal -- The heaviest of the alloying-forming metals which react readily with dithizone

Hot topping -- A practice to provide feed metal to the ingot as it contracts on solidification to avoid top end voids

Leached and dried sponge -- Sponge which is subsequently leached with an aqueous solution and then dried; usually applied to titanium

Liquidus temperature -- The maximum temperature at which crystals can exist with liquid at a specified pressure

Monel metal -- A nickel base alloy with compositions ranging in weight percent from 63-67 Ni, 30 Cu, 3 Al, and/or 1.5 to 4 Si, and/or Mn

Primary melting -- The first melting operation to produce a homogeneous alloy with the composition required in the final product

Pyrometallurgical refining -- Refining where heat is used to accomplish specific purification reactions

Radial segregation -- Dark segregation streaks arranged in a radial pattern predominantly in mid-radius regions

Rare earth fluorides -- Lanthanum (atomic no. 57-71) and yttrium (atomic no. 39) fluorides

Reactive metal -- A metal which is readily susceptible to chemical change, especially one which readily oxidizes at high temperatures

Refractory metal -- A metal with a very high melting point which resists the effects of heat and which is often difficult to work at elevated temperatures

Secondary melting -- A second melting (re-melting) operation to further refine an alloy and impart specific characteristics on the ingot without significantly changing its composition

Slag cap -- The thickness of the slag cover on top of the electroslag ingot representing the primary resistance element in the circuitry

Sponge -- A form of metal characterized by a porous condition which is the result of the decomposition or reduction of a compound without fusion

Tree-ring pattern -- Series of concentric bands of light and dark appearance which are most pronounced in billet products

Upset forging -- A means of increasing the diameter of a bar during forging by striking it on the end

Vacuum distilled sponge -- Sponge which has been purified in a vacuum to remove excess magnesium and chlorides; usually applied to titanium

VAR -- Vacuum arc remelting

VOD -- Vacuum oxygen decarburizing

Yield -- The percentage of useable material obtained in processing of ingots to semi-finished or finished products; sometimes also called "recovery"

APPENDIX B

ELECTROSLAG FURNACE INSTALLATIONS

The location, size, type, and estimated annual production capacities of ESR furnaces in the world are listed in Table 31 by country (alphabetically). Included are installed furnaces, known furnaces that will be installed during 1975, and known vacuum arc remelt furnaces that have been converted for DC electroslag melting. The maximum capacities are listed (i.e., ingot size, current rating, etc.); however, the furnaces may not be operating at this capacity. For instance, many of the furnaces are equipped with several sizes of static molds, and the largest size ingot is produced only occasionally. ESR furnace capacity is a function of furnace design and operating parameters that include:

1) Number of melt stations and furnace heads, which influence turn-around time between melts
2) KW available at the melt zone, which is influenced by electrical circuit design
3) Type of molds used -- fixed or collar
4) Molten slag or cold starts
5) Electrode/ingot area ratio
6) Slag composition (resistivity) and slag depth
7) Ingot diameter and length
8) Segregation sensitivity of the alloy being melted (for sensitive grades, melt rates too slow or too fast can cause segregation problems).

The annual production estimates are based on the maximum current ratings of the power supplies and on the following assumptions:

1) Seventy-five percent of the rated maximum current is available at the melt station at a potential across the slag pool of 40 V
2) Melt power (kW) is the simple product of $V \cdot A$
3) The specific power consumption is 1200 kWhr/ton, including a startup and hot topping
4) The furnace is operated 52 weeks per year, at 21 turns per week, and 85 percent utilization (including turn-around time and hot-topping); this is about 7426 operating hours per year.

TABLE 31 World Electroslag Furnace Installations.

Location	Design and Year	Furnace Type*	Ingot Cross Section (inches)	Ingot Weight (pounds max.)	Power Supply	Amps (max.)	Estimated Capacity (tons/yr)	Alloys Produced
AUSTRIA								
Boehler Brothers & Co., Ltd. Kapfenberg	Boehler 1967	II-C1 Round	40 dia.	48,000	1 ∅ AC 50 Hz	13,000	2,400	Tool and die steels High strength steels Stainless steels
	Boehler 1969	II-C1 Round	22 dia.	23,000	1 ∅ AC 50 Hz	18,000	3,340	Tool and die steels High strength steels Stainless steels
BELGIUM								
Cockerill Ougree	Boehler 1972	--	40 dia.	50,000	1 ∅ AC 3-5 Hz	30,000	5,570	Roll steels
	Boehler 1974	II-C1 Round	40 dia.	50,000	1 ∅ AC 3-5 Hz	30,000	5,570	Roll steels
BRAZIL								
Electrometal-Acos Finos Sumare	CONSARC 1973	I-A2 Round Slab	32 dia. 18 x 36	20,000	1 ∅ AC 60 Hz	22,300	4,140	Tool and die steels Stainless steels Bearing steels
Institute de Pesquisas Tecnologicas Sao Paulo	CONSARC 1975	I-A1 Round	6.5 dia.	225	1 ∅ AC 60 Hz	3,000	--	Research
FRANCE								
Compagnie des Ateliers et Forges de la Loire (CAFL) Firmini	USSR 1965	I-A1 Round Slab	20 dia. 14 x 22	5,500	1 ∅ AC 50 Hz	14,000	2,600	Low alloy steels High strength steels Stainless, high speed and super alloys
Aubert et Duval	CONSARC 1973	II-A2	33 dia.	30,000	1 ∅ AC 50 Hz	30,000	5,570	Alloy steels Tool steels Stainless steels
Commentryenne des Aciers Fins	Boehler-Heurty 1970	II-C1	22 dia.	--	1 ∅ AC	--	--	Tool and die steels High strength steels Specialty alloys
Creusot-Loire (CAFL) Firmini	USSR 1969	II-A1 Bifilar Monofilar	--	--	1 ∅ AC	--	--	Low alloy steels High strength steels Stainless steels Tool steels

* Roman numerals indicate the number of electrode suspension heads. The capital letter-Arabic number combinations indicate the types of melt station used and the number of each type (A = full-length static mold; B = ingot withdrawal; C = moving mold).

TABLE 31 World Electroslag Furnace Installations (continued).

Location	Design and Year	Furnace Type*	Ingot Cross Section (inches)	Ingot Weight (pounds max.)	Power Supply	Amps (max.)	Estimated Capacity (tons/yr)	Alloys Produced
INDIA								
Tata Jamshedpur (6 furnaces)	Self-built 1974	I-B1 Square	10 x 10	2,800	--	--	--	Tool steels
ITALY								
Societa Nationale Cogne Aosta	Birlec 1967	III-B1 Square	32.4 x 32.4	12,000	1 ø AC 3 ø AC	--	--	Constructional steels Tool and die steels Stainless steels
JAPAN								
Daido Steel Co., Ltd.	Self-built 1971	I-A1 Round	22 dia.	5,700	1 ø AC 60 Hz	15,300	2,750	Stainless heat resistant Tool steels Constructional steels
Hitachi Metals, Ltd.	Self-built 1967	I-A1 Round	13.5 dia.	2,200	60 Hz	7,500	1,350	Tool steels Nickel-base alloys
	Self-built 1967	I-A1 Round	28.5 dia.	13,000	60 Hz	15,000	2,400	Tool steels Nickel-base alloys
	Self-built 1968	I-A1 Round	32 dia.	22,000	60 Hz	20,000	3,600	Tool steels Nickel-base alloys
	Self-built 1971	I-A1 Round	13.5 dia.	2,200	60 Hz	10,000	1,800	Tool steels Nickel-base alloys
Japan Special Steel Co., Ltd.	Self-built 1966	I-B1 Round	14 dia.	2,400	60 Hz	12,500	2,250	Nickel-base alloys Tool steels Stainless steels
	Self-built 1968	I-A1 Round	22 dia.	8,800	1 ø AC 60 Hz	20,000	3,600	Nickel-base alloys Tool steels Stainless steels
Japan Steel Works, Ltd.	Ingot withdrawal 1970	I-B1 Round	11 dia.	1,100	60 Hz	5,500	1,000	--

* Roman numerals indicate the number of electrode suspension heads. The capital letter-Arabic number combinations indicate the types of melt station used and the number of each type (A = full-length static mold; B = ingot withdrawal; C = moving mold).

TABLE 31 World Electroslag Furnace Installations (continued).

Location	Design and Year	Furnace Type*	Ingot Cross Section (inches)	Ingot Weight (pounds max.)	Power Supply	Amps (max.)	Estimated Capacity (tons/yr)	Alloys Produced
JAPAN (continued)								
Kanto Special Steel Works, Ltd.	Self-built 1971	I-A1 Round	12 dia.	650	1 ø AC 60 Hz	6,000	1,080	Roll steels
Kobe Steel Ltd.	Self-built 1970	I-A1 Round	60 dia.	110,000	1 ø AC 60 Hz	50,000	9,000	Forging alloy
Mitsubishi Heavy Industries Ltd.	Self-built 1970	II-B1 Hollow	20 O.D. x 19 I.D. x 400 long	6,600	3 ø AC 60 Hz	3,000	540	Hollows of heat-resistant and stainless alloys
	Self-built 1971	II-B1 Hollow	10 O.D. x 9 I.D. x 520 long	--	3 ø AC 60 Hz	3,000	540	Hollows of heat-resistant and stainless alloys
Mitsubishi Steel Manufacturing Co., Ltd.	Self-built 1969	II Round Square	13.5 dia. or square	1,300	1 ø AC 60 Hz	9,000	1,620	Constructional steels
	Leybold-Heraeus-Daido 1970	II-C1 Round	29 dia.	17,000	DC	20,000 (est.)	3,600	Constructional steels
Mitsubishi Metals and Mining Co., Ltd.	Self-built 1969	I-A1 Round	Unknown (approx. 20 dia.)	6,600	60 Hz	19,000	3,400	Constructional steels
Nippon Kinzoku Co., Ltd.	Self-built 1971	I-A1 Round	6.5 dia.	330	1 ø AC 60 Hz	10,000	1,800	Experimental
Nippon Koshuha Steel Co., Ltd.	Self-built 1968	I-A1 Round	14.5 dia.	2,200	1 ø AC 60 Hz	7,500	7,500	--
Nippon Steel Corp.	CONSARC 1971	II-B2 Round Slab	14 dia. 18 x 48 x 80	15,500	1 ø AC 60 Hz	30,000	5,400	Constructional steels
	USSR Bifilar (Static & ingot withdrawal)	I-C1 Round Slab	--	60,000	1 ø AC 60 Hz	>50,000	9,000	Constructional steels
Pacific Metals Co., Ltd.	Self-built 1972	I-A1 Round	20 dia.	4,800	1 ø AC 60 Hz	10,000	1,800	--

* Roman numerals indicate the number of electrode suspension heads. The capital letter-Arabic number combinations indicate the types of melt station used and the number of each type (A = full-length static mold; B = ingot withdrawal; C = moving mold).

TABLE 31 World Electroslag Furnace Installations (continued).

Location	Design and Year	Furnace Type*	Ingot Cross Section (inches)	Ingot Weight (pounds max.)	Power Supply	Amps (max.)	Estimated Capacity (tons/yr)	Alloys Produced
JAPAN (continued)								
Riken Piston Co., Ltd.	Self-built 1957	I-A1 Round	14 dia.	1,100	60 Hz	10,000	1,800	Electric resistant alloys
	Self-built 1959	I-A1 Slab	Rectangular	11,400	60 Hz	10,000	1,800	Electric resistant alloys
Sanyo Special Steel Co., Ltd.	USSR 1972	I-A1 Round Square	17 x 17	4,400	1 ø AC 60 Hz	15,500	2,800	Specialty steels and alloys
Tokushu Seiko Co., Ltd.	Self-built 1966	I-A1 Round	11.5 dia.	1,100	1 ø AC 60 Hz	10,000	1,800	Nickel-base alloys
	Self-built 1968	I-A1 Round	15 x 15	2,400	1 ø AC 60 Hz	14,000	2,500	Nickel-base alloys
	Self-built 1973	I-A1 Round	12 dia.	1,200	1 ø AC 60 Hz	7,000	1,250	Nickel-base alloys
ULVAC Corp.	USSR 1973	Round	18 dia.	5,500	60 Hz	14,000	2,500	Pilot Furnace
LUXEMBURG								
ARBED Dommeldange	CONSARC 1971	II-A2 Round	30 dia.	24,000	1 ø AC 50 Hz	25,000	4,640	Tool and die steels Roll steels High strength steels Specialty steels
SPAIN								
Astilleros Espanole Reinosa	CONSARC 1975	II-A2	40 dia.	40,000	1 ø AC 50 Hz	40,000	7,400	Constructional steels
SWEDEN								
Avesta Jernwerke AB, Avesta	USSR 1971	I-A1 Round Slab	--	18,000	1 ø AC	--	--	Stainless steels
Bofors AB Bofors	Boehler 1970	II-C1 Round	20 dia.	14,000	1 ø AC	10,000	1,850	Die steels Constructional steels Stainless steels Heat resistant steels
	Boehler 1970	II-C1 Round	40 dia.	60,000	1 ø AC	20,000	3,700	Die steels Constructional steels Stainless steels Heat resistant steels

* Roman numerals indicate the number of electrode suspension heads. The capital letter-Arabic number combinations indicate the types of melt station used and the number of each type (A = full-length static mold; B = ingot withdrawal; C = moving mold).

TABLE 31 World Electroslag Furnace Installations (continued).

Location	Design and Year	Furnace Type*	Ingot Cross Section (inches)	Ingot Weight (pounds max.)	Power Supply	Amps (max.)	Estimated Capacity (tons/yr)	Alloys Produced
SWEDEN (continued)								
Fagersta Bruks AB Surahammar	GWB 1968	III-B1 Round	--	--	1 ø AC 3 ø AC	--	--	Tool steels
Sandviken	Leybold Heraeus	Converted VAR	20 dia.	--	DC	15,000	2,780	Stainless steels
Stora Kopparberg Soderfors	Birlec	III-C1	24 dia.	--	1 ø AC	--	--	Tool steels
Uddeholm	Boehler 1969	II-C1	32 dia.	--	1 ø AC	--	--	Tool and die steels
	Boehler 1970	II-C1	32 dia.	--	1 ø AC	--	--	Tool and die steels
	Boehler 1972	II-C1	32 dia.	--	1 ø AC	--	--	Tool and die steels
	Boehler 1975	II-C1	32 dia.	--	1 ø AC	--	--	Tool and die steels
UNITED KINGDOM								
Birlec Ltd.	Birlec 1963	III-B1	14 dia. 30 dia.	--	700 KVA 1 and 3 ø	14,000	2,600	Experimental
British Steel Corp.	Birlec 1975	III-AB1	--	--	3 ø AC	--	--	Constructional steels
Brown Bayley Steels	Birlec 1967	I-A1	10 dia.	--	1 ø AC 330 KVA	6,400	1,140	Tool steels
English Steel Forge & Engineering Corp.	ESC 1966	I-A1	36 dia.	--	1 ø 1200 KVA	24,000	4,440	Constructional steels Research
Firth Brown	Firth Brown 1963	I-A1	24 dia. modified to 40 dia.	--	1 ø 600 KVA	12,000	2,220	Roll steels Constructional steels Stainless steels Die steels Tool steels
	Birlec 1973	I-A1	26 dia.	14,000	1 ø 3 ø	--	--	Roll steels Die steels Constructional steels
Henry Wiggin	Birlec 1967	III-B1	16 dia. 48 dia.	--	1 ø	25,000	4,640	Super alloys Heat resistant alloys
	Birlec 1970	III-B1	25 dia. 43 dia.	1 ø 10,000 3 ø 30,000	1 ø 3 ø 3 x 2000 KVA	40,000	7,400	Super alloys Heat resistant alloys

* Roman numerals indicate the number of electrode suspension heads. The capital letter-Arabic number combinations indicate the types of melt station used and the number of each type (A = full-length static mold; B = ingot withdrawal; C = moving mold).

TABLE 31 World Electroslag Furnace Installations (continued).

Location	Design and Year	Furnace Type*	Ingot Cross Section (inches)	Ingot Weight (pounds max.)	Power Supply	Amps (max.)	Estimated Capacity (tons/yr)	Alloys Produced
UNITED KINGDOM (continued)								
Jones & Colver	Birlec 1967	I-A1	10 dia.	--	1 ø AC 250 KVA	5,000	920	Tool steels
Joseph Gillett	Rubery Owen 1966	I-A1	10 dia.	--	1 ø 200 KVA	4,000	740	Tool steels Constructional steels
Kayser Ellison	Birlec 1967	III-B1	11 dia. 1 ø 34 dia. 3 ø	--	1 & 3 ø 750 KVA	15,000	2,775	Tool steels
Osborn Steels Bradford Works	Birlec 1967	III-B1	14 dia. 1 ø 36 dia. 3 ø	--	1 & 3 ø AC 1080 KVA	--	4,000	Tool steels
Richard W. Carr	Birlec 1968	III-B1	9 dia. 1 ø 24 dia. 3 ø	--	1 & 3 ø AC 540 KVA	10,800	1,900	Tool steels
Union Carbide Glossup	Birlec 1970	III-B1	22 dia. 1 ø 36 dia. 3 ø	9,000 14,000	1 & 3 ø AC 2 x 1600 KVA	32,000	5,900	Super alloys Heat resistant steels
William Oxley	Wild Barfield 1968	III-B1	36 dia. 3 ø		1 & 3 ø AC 1350 KVA	27,000	5,000	Stainless steels
UNITED STATES								
Allegheny Ludlum Special Metals Div. New Hartford, NY	CONSARC 1968	I-A2 Round	8 dia.	300	1 ø AC 60 Hz	4,000	670	Nickel-base alloys Specialty steels
Allegheny Ludlum Watervliet, NY	Zak 1969	I-A1 Round	20 dia.	7,000	DC ED +	18,000	3,340	High-speed tool steels
	Zak 1969	I-A1 Round	20 dia.	7,000	DC ED +	18,000	3,340	High-speed tool steels
Babcock & Wilcox Beaver Falls, PA	CONSARC 1973	II-A2 Round	30 dia.	25,000	1 ø AC 60 Hz	20,000	3,700	High-strength steels Specialty steels High-temperature alloys
Bethlehem Steel Bethlehem, PA	Leybold-Heraeus 1975	II-B1 Round Slab	26 dia. 13 x 17	20,000	1 ø AC 3-10 Hz	18,000	3,340	Tool steels Specialty steels
	Leybold-Heraeus 1975	II-B2 Round	60 dia.	150,000	1 ø AC 3-10 Hz	55,000	10,200	Tool steels Die steels Low-alloy steels
Cabot Corp. Pampa, TX	CONSARC 1975	II-A2-B1 Round Hollow	30 dia.	30,000	1 ø AC 60 Hz	30,000	5,570	Low-alloy steels

* Roman numerals indicate the number of electrode suspension heads. The capital letter-Arabic number combinations indicate the types of melt station used and the number of each type (A = full-length static mold; B = ingot withdrawal; C = moving mold).

TABLE 31 World Electroslag Furnace Installations (continued).

Location	Design and Year	Furnace Type*	Ingot Cross Section (inches)	Ingot Weight (pounds max.)	Power Supply	Amps (max.)	Estimated Capacity (tons/yr)	Alloys Produced
UNITED STATES (continued)								
Cabot Corporation Stellite Division Kokomo, IN	CONSARC 1967	II-A2 Round	33 dia.	35,000	1 ø AC 60 Hz	25,000	4,640	Nickel-base alloys Cobalt-base alloys Heat-resistant alloys
	CONSARC 1968	I-A2 Round	6 dia.	80	1 ø AC 60 Hz	3,300	--	Research
	CONSARC 1968	II-A2 Round	33 dia.	35,000	1 ø AC 60 Hz	25,000	4,640	Nickel-base alloys Cobalt-base alloys Heat-resistant alloys
	CONSARC 1971	II-A2-C1 Round Hollow	24 dia.	20,000	1 ø AC	25,000	4,640	Nickel-base alloys Cobalt-base alloys Heat-resistant alloys
	Leybold-Heraeus 1975	II-A&B	30 dia.	30,000	1 ø AC 60 Hz	20,000	4,640	Nickel-base alloys Cobalt-base alloys Heat-resistant alloys
Carpenter Technology Reading, PA	NRC 1967	I-A1 Round	20 dia.	5,000	DC	18,000	3,340	Tool and die steels Specialty steels Low alloy steels Stainless steels
	NRC 1967	I-A1 Round	20 dia.	5,000	DC	18,000	3,340	Tool and die steels Specialty steels Low alloy steels Stainless steels
	CONSARC 1968	I-A1 Round	8 dia.	300	1 ø AC 60 Hz	4,000	--	Research
	Boehler 1972	II-C1 Round	30 dia.	5,000	1 ø AC 60 Hz	16,000	2,970	Tool and die steels Specialty steels Stainless steels
	Boehler 1972	II-C1 Round	30 dia.	5,000	1 ø AC 60 Hz	16,000	2,970	Tool and die steels Specialty steels Stainless steels
	Boehler 1974	II-C1 Round	40 dia.	50,000	1 ø AC 3-5 Hz	30,000	5,570	Tool and die steels Specialty steels Stainless steels
CONSARC Corp. Rancocas, NJ	CONSARC 1967	Variable (research)	Variable	30,000	AC & DC	60,000	--	Research
Crucible Steel Syracuse, NY	CONSARC	I-A2 Round (may phase out ESR)	30 dia.	27,000	DC ED -	20,000	3,700	Tool and die steels

* Roman numerals indicate the number of electrode suspension heads. The capital letter-Arabic number combinations indicate the types of melt station used and the number of each type (A = full-length static mold; B = ingot withdrawal; C = moving mold).

TABLE 31 World Electroslag Furnace Installations (continued).

Location	Design and Year	Furnace Type*	Ingot Cross Section (inches)	Ingot Weight (pounds max.)	Power Supply	Amps (max.)	Estimated Capacity (tons/yr)	Alloys Produced
UNITED STATES (continued)								
Electro Slag Institute Washington, PA	Self-built 1968	Round Slab	16 dia. 7 x 14	2,000	1 ∅ AC 60 Hz	17,800	--	Research
	Self-built 1968	Round Slab	22 dia. 12 x 36	6,000	1 ∅ AC 60 Hz	17,800	--	Research
International Nickel Co. Burnaugh, KY	CONSARC 1971	I-A2 Slab	12 x 53	20,000	1 ∅ AC 60 Hz	20,000	3,700	Nickel-base alloys Cobalt-base alloys Other heat-resistant alloys
	CONSARC 1971	I-A2 Slab	12 x 53	20,000	1 ∅ AC 60 Hz	20,000	3,700	Nickel-base alloys Cobalt-base alloys Other heat-resistant alloys
International Nickel Co. Huntington, WV	CONSARC 1969	II-A2 Round Slab	40 dia.	40,000	1 ∅ AC 60 Hz	30,000	5,570	Nickel-base alloys Cobalt-base alloys Other heat-resistant alloys
Latrobe Steel Co. Latrobe, PA	CONSARC 1968	I-A2 Round	24 dia.	13,000	DC ED +	18,000	3,340	Tool and die steels Specialty steels Stainless steels
Lukens Steel Co. Coatesville, PA	CONSARC 1970	II-A2 Slab	30 x 80	60,000	1 ∅ AC 60 Hz	60,000	11,100	Die steels High-strength steels Plate steels
Simonds Steel Co. Lockport, NY	CONSARC 1969	I-A2 Round (primarily VAR)	33 dia.	33,000	DC	20,000	3,700	Tool and die steels Stainless steels Specialty steels High-strength steels
	CONSARC 1969	II-A2 Round	33 dia.	33,000	1 ∅ AC 60 Hz	25,000	4,640	Tool and die steels Stainless steels Specialty steels High-strength steels
	CONSARC 1974	II-A2 Round	33 dia.	33,000	1 ∅ AC 60 Hz	25,000	4,640	Tool and die steels Stainless steels Specialty steels High-strength steels
Teledyne Allvac Monroe, NC	Leybold-Heraeus 1973	II-B2 Round	20 dia.	15,000	1 ∅ AC 60 Hz	15,000	2,780	Superalloys
Teledyne Vasco Latrobe, PA	CONSARC 1966	I-A2 Round	33 dia.	33,000	DC	20,000	3,700	Tool and die steels High-strength steels Superalloys

* Roman numerals indicate the number of electrode suspension heads. The capital letter-Arabic number combinations indicate the types of melt station used and the number of each type (A = full-length static mold; B = ingot withdrawal; C = moving mold).

TABLE 31 World Electroslag Furnace Installations (continued).

Location	Design and Year	Furnace Type*	Ingot Cross Section (inches)	Ingot Weight (pounds max.)	Power Supply	Amps (max.)	Estimated Capacity (tons/yr)	Alloys Produced
UNITED STATES (continued)								
Union Carbide Nuclear Division Oak Ridge, TN	CONSARC 1970	I-A2 Round	20 dia.	15,000	DC	15,000	--	Classified nuclear materials
	CONSARC 1971	I-A2 Round	8 dia.	300	DC	10,000	--	Research
Union Electric Co. Burgettstown, PA	Leybold-Heraeus 1967	I-A1 Round Slab (quick change)	30 dia.	24,000	DC	22,500	4,170	Forged roll compositions
Universal Cyclops Bridgeville, PA	CONSARC 1970	II-A2-B1 Round Slab	24 dia.	15,000	DC and 1 ø AC 60 Hz	15,000	2,780	Tool and die steels Nickel-base alloys Cobalt-base alloys Specialty steels
	CONSARC 1975	I-C1 Round Slab	30 dia.	24,000	1 ø AC 60 Hz	30,000	5,570	Tool and die steels Nickel-base alloys Cobalt-base alloys Specialty steels
U.S. Army Materials and Mechanics Research Center Watertown, MA	CONSARC 1974	I-A1 Round Slab	8 x 12	500	1 ø AC 60 Hz	8,000	--	Research
U.S. Bureau of Mines Albany, OR	Self-built 1965	I-A1 Round Slab	2.5-6.0 dia. 3.6 x 6.9	160	1 ø AC 60 Hz or DC	4,500 7,000∓	--	Research
	Self-built 1967	I-A1 Round	10 dia.	900	DC	7,000∓	--	Research
	Self-built 1975	I-B1 Round	4 dia.	100	1 ø AC 60 Hz	2,500	--	Research
Watervliet Arsenal Watervliet, NY	CONSARC 1974	I-C1 Round Hollow	8 dia.	300	1 ø AC 60 Hz	8,000	--	Research
Wilber B. Driver Co. Newark, NJ	CONSARC 1975	II-A2 Round	20 dia.	6,000	1 ø AC 60 Hz	15,000	2,780	Stainless steels Controlled expansion alloys

* Roman numerals indicate the number of electrode suspension heads. The capital letter-Arabic number combinations indicate the types of melt station used and the number of each type (A = full-length static mold; B = ingot withdrawal; C = moving mold).

∓ Actual usage; maximum DC rating is 18,000 amps at 30 volts.

TABLE 31 World Electroslag Furnace Installations (continued).

Location	Design and Year	Furnace Type*	Ingot Cross Section (inches)	Ingot Weight (pounds max.)	Power Supply	Amps (max.)	Estimated Capacity (tons/yr)	Alloys Produced
WEST GERMANY								
Boehler Brothers and Co., Ltd. Dusseldorf	Boehler 1969	II-C1	25 dia.	16,000	1 ø AC 50 Hz	15,000	2,780	High-speed, stainless constructional steels
Deutsch Edelstahlwerke AG (DEW), Krefeld	CONSARC 1970	I-A2	20 dia.	10,000	1 ø AC 50 Hz	15,000	2,780	Tool and die steels High-strength steels Stainless steels
Edelstahlwerke Witten, AG Witten	Leybold-Heraeus 1969	II-A2	26 x 26	20,000	1 ø AC 50 Hz	20,000	3,700	Stainless, high-speed constructional steels
Kloeckner, Osnabruck	Kloeckner 1974	I-A1	52 dia.	--	1 ø AC 50 Hz	--	--	Low alloy steels Roll steels CZR process
Krupp	Boehler 1972	II-C1	40 dia.	--	AC (low freq.)	--	--	Constructional steels
Rheinstahl Huettenwerke AG Hattingen-Ruhr	Rheinstahl 1966	I-A1	12 dia.	--	1 ø AC 50 Hz	--	--	Experimental alloys
	Rheinstahl 1967	--	40 dia.	9,000	3 ø AC 50 Hz	--	--	Constructional steels Stainless steels
Rochling, Volklingen	Leybold-Heraeus 1971	IV-B1	92 dia.	350,000	AC (low freq.)	100,000	18,500	Constructional steels Rotor steels
Stahlwerke Sudwestfahlen, Geisweid	Birlec, Junkers 1969	III-A1	32 x 32	--	1 ø AC 3 ø AC	--	--	Tool and die steels Constructional steels Stainless steels Heat-resistant steels

* Roman numerals indicate the number of electrode suspension heads. The capital letter-Arabic number combinations indicate the types of melt station used and the number of each type (A = full-length static mold; B = ingot withdrawal; C = moving mold).

APPENDIX C

REFERENCES AND BIBLIOGRAPHY

Andreev, A.D., Mikheichev, L.A., Kurdyumov, A.V., and Bozhenek, I.V. "Electric Flux Refining of Aluminum Alloys." Soviet Journal of Non-Ferrous Metals 8 (1967):97-100.

Antoine, L., Jallas, P., Desolneux, J.P., and Boucher, A. "Operating Results of an Industrial Electroslag Consumable Electrode Remelting Plant at CAFL." In Proc. First International Symposium on Electroslag Consumable Electrode Remelting and Casting Technology, Vol. I. Pittsburgh: Mellon Institute, 1967.

Armantrout, C.E., Dunham, J.T., and Beall, R.A. "Properties of Wrought Shapes Formed from Electroslag-Melted Titanium." In The Science, Technology, and Application of Titanium, pp. 67-74. New York: Pergamon Press, 1970.

Armantrout, C.E., and Nafziger, R.H. "Development of Electroslag Melting Technique for Titanium; Selected Properties of Fabricated Material." Trans. American Foundrymen's Society 77(1969): 353-9.

Artamonov, V.L., ed. Scientific Bibliography--Special Electrometallurgy. Kiev, 1972.

_____. Collection of Annotations of Foreign Patents Relating to Electroslag Refining. Report No. JPRS 57283. Arlington, Virginia: Joint Publications Research Service, 1972.

Arwidson, S.G. "Technico-Economic Appraisal of a Large Tonnage ESR Installation." In Electroslag Refining, pp. 157-62. London: Iron and Steel Institute, 1973.

Asada, C., and Adachi, T. "On Plasma Induction Melting." In Proc. Third International Symposium on Electroslag and Other Special Melting Technology. Pittsburgh: Mellon Institute, 1971.

Ausmus, S.L., and Beall, R.A. "Electroslag Melting of Titanium Slabs." In Trans. International Vacuum Metallurgy Conference, pp. 675-94. New York: American Vacuum Society, 1968.

Bagshaw, T., et al. "Application of ESR to Roll Development." In Electroslag Refining, pp. 126-34. London: Iron and Steel Institute, 1973.

Baird, R.J. "High Voltage Arc Plasma Generator." U.S. Patent No. 3,194,941 (July 13, 1965).

Beall, R.A., Calvert, E.D., Clites, P.G., and Dunham, J.T. "Electroslag Melting of Titanium and Molybdenum." In Proc. First International Symposium on Electroslag Consumable Electrode Remelting and Casting Technology, Vol. I. Pittsburgh: Mellon Institute, 1967.

Beall, R.A., Clites, P.G., Nafziger, R.H., and Dunham, J.T. Titanium Melting by the Electroslag Process. Report No. USBM-RC-1351. Washington, D.C.: U.S. Bureau of Mines, 1969.

Belous, G. Avton. Svarka 65(1958): 32.

Bhat, G.K. "Innovations in Plasma Heat Applications." In Proc. Fifth International Symposium on Electroslag and Other Special Melting Technology. Pittsburgh: Mellon Institute, 1974.

_____. "Manufacture of Shaped Castings Through the Electroslag Remelting Process." In Proc. Fourth International Symposium on Electroslag Remelting Processes. Tokyo: The Iron and Steel Institute of Japan, 1973.

_____. "Manufacture of Shaped Castings Through the Electroslag Remelting Process." In Reports of the Symposium on Special Electrometallurgy. Kiev: Soviet Academy of Sciences, 1972.

_____. "New Developments in Plasma Arc Melting." Journal of Vacuum Science Technology 9(November/December, 1972).

_____. "Techniques of Electroslag Remelting of Hollow Ingots." In Proc. Third International Symposium on Electroslag and Other Special Melting Technology. Pittsburgh: Mellon Institute, 1971.

_____. A Manufacturing Program for the Electroslag Melting and Casting of Materials." Report No. AFML-TR-71-162. Wright-Patterson Air Force Base, Ohio: Air Force Materials Laboratory, 1971.

Boucher, A. Bulletin of the Centre des Etudes Metallurgie 12(1972): 229.

Cadden, J.L., Jessen, N.C., and Lewis, P.S. "Melting of Uranium Alloys." Paper presented at Physical Metallurgy of Uranium Alloys Conference, Vail, Colorado, February 1974. (Preprint Y-DA-5323, Oak Ridge Y-12 Plant, 1973.)

Calvert, E.D., Beall, R.A., and Kato, H. "Electroslag-Melted Molybdenum." Journal of Less Common Metals 23(1971): 129-51.

Cameron, J., Etienne, M., and Mitchell, A. "Some Electrical Characteristics of a DC Electroslag Unit." Metallurgical Transactions 1(1970): 1839-44.

Chernega, D. Avton. Svarka 66(1959): 99.

Choudhury, A.; Jauch, R.; Hinze, H.; and Scheidig, H. "Production of Heavy Ingots: A New Era of ESR." Proc. Third International Symposium on Electroslag and Other Special Melting Technology, Vol. II, p. 159. Pittsburgh: Mellon Institute, 1971.

Chulkov, V.S., Mikheichev, L.A., and Kurdyumov, A.V. "The Effects of Flux Composition and State of Aggregation on the Micro-Porosity and Gas Saturation of Billets During Electroslag Refining." Tekhnol. legkikh splavov, nauckno-tekhn. Byal. VIISa 4(1970): 34-6.

Cook, R.L., Reese, G.W., Jr., and Gadsby, P.M. "Electroslag Remelting of Nickel Base Alloys." In Proc. Third International Symposium on Electroslag and Other Special Melting Technology, p. 142. Pittsburgh: Mellon Institute, 1971.

Cooper, L.R. Method of Producing Large Steel Ingots. U.S. Patent No. 3,603,374, Heppenstall Company (September 7, 1971).

_____. Production of Large Steel Ingots with Consumable Vacuum Arc Hot Tops. U.S. Patent No. 3,696,859, Heppenstall Company (October 10, 1972).

Cooper, L.R., Mogendorf, W., and Heymann, H. "Central Zone Remelting of Ingots Using an Electroslag Process." In Proc. Fifth International Symposium on Electroslag and Other Special Melting Technology, Vol. I, pp. 202-38. Pittsburgh: Mellon Institute, 1974.

Cremisio, R.S., and Zak, E.D. "Considerations of Mold Design Parameters and ESR Production Technology." In Proc. Fourth International Symposium on Electroslag Remelting Processes, pp. 137-48. Tokyo: The Iron and Steel Institute of Japan, 1973.

Descamps, T., and Etienne, M. "Production of Larger Ingots by Continuous Electroslag Powder Melting." In Electroslag Refining, pp. 150-4. London: Iron and Steel Institute, 1973.

Dewsnap, P., and Schlatter, R. "Process and Product Characteristics of DC Electroslag Remelting of Alloy Steel." In *Proc. Fifth International Symposium on Electroslag and Other Special Melting Technology*, pp. 91-114. Pittsburgh: Mellon Institute, 1974.

Dorschu, K.E. "Production of Special-Steel Billets by Continuous Electroslag Powder Melting." In *Electroslag Refining*, pp. 145-9. London: Iron and Steel Institute, 1973.

Duckworth, W.E., and Hoyle, G. *ESR Plant Construction*. London: Chapman and Hall, Ltd., 1969.

"Electroslag Remelting of Steel in France." *Journal of Metals* (February, 1966).

Elliott, C.F., Vorberger, J.W., Staton, M.G., and Mills, G.L. "Superalloy Melting Using an Ingot Bottom Withdrawal AC ESR Furnace." In *Proc. Fourth International Symposium on Electroslag Remelting Processes*, pp. 298-310. Tokyo: The Iron and Steel Institute of Japan, 1973.

Etienne, M., and Mitchell, A. "Oxidative Losses of Low Levels of Titanium During Electroslag Remelting." *Proc. Second International Symposium on Electroslag and Other Special Melting Technology*, Vol. II. Pittsburgh: Mellon Institute, 1969.

Fiedler, H., Lachner, W., Belka, G., and Muller, F. "Results of Plasma Melting of Steel." In *Proc. Fifth International Symposium on Electroslag and Other Special Melting Technology*. Pittsburgh: Mellon Institute, 1974.

Gage, R.M. *Arc Torch and Process*. U.S. Patent No. 2,806,124 (September 10, 1957).

Gill, L.L., and Harris, K. "Development of a VIM plus ESR Process Route for Direct Forging of Components from Ingot in Nickel-Base Precipitation Hardened Super Alloys." In *Electroslag Refining*, pp. 89-101. London: Iron and Steel Institute, 1973.

Gurevich, S.M., Didkovsku, V.P., Novikov, Yu. K., Tilorikyan, B.K., Zasetsku, G.F., Kravchenko, V.F., and Novikona, A.A. "The Electroslag Melting of Titanium Alloy Ingots." *Automatic Welding* 16 (nos. 4 and 19, 1963): 20-6; 33-8.

Hoffman, A.O., Battelle-Columbus Laboratories, Columbus, Ohio. Personal Communication, September 1974.

Holzgruber, W. "Ergebnisse aus dem Betrieb einer Grosstechnischen ESR Anlage." Paper No. 139 presented at Electro-Heat Conference, Brighton, England, 1968.

Holzgruber, W., Kroneis, M., and Schneidhofer, A. "The Production of Ingots up to 50,000 Pounds in the Boehler ESR Unit." In Proc. Second International Symposium on Electroslag and Other Special Melting Technologies, Vol. I. Pittsburgh: Mellon Institute, 1969.

Holzgruber, W., Machner, P., and Ploeckinger, E. "Mechanism and Results of Improved Metal Refinement Using Superimposed DC Electrolytic Refining During Electroslag Remelting." In Trans. International Vacuum Metallurgy Conference, Anaheim, California, pp. 415-30. Columbus, Ohio: Battelle Memorial Institute, 1969.

Holzgruber, W., and Ploeckinger, E. Stahl u. Eisen 88, 12(1968): 638-48.

Hopkins, R.K. "The Electric Ingot Process." In Proc. Electric Furnace Steel Conference, pp. 91-105. New York: American Institute of Mining, Metallurgical and Petroleum Engineers, 1948.

_____. Manufacture of Metal Articles. U.S. Patent No. 2,191,475, M.W. Kellogg Company (1940).

_____. Manufacture of Alloy Ingots. U.S. Patent No. 2,191,479, M.W. Kellogg Company (February 27, 1940).

Hoyle, G. "Production of Small Ingots and Hollows by ESR." In Electroslag Refining, pp. 136-44. London: The Iron and Steel Institute, 1973.

Jackson, R.O., Mitchell, A., and Luchok, J. "An Examination of Electrode-Change Practice in Electroslag Melting." Journal of Vacuum Science Technology 9 (November/December, 1972).

Joint Publications Research Service. Electroslag Casting. Report No. JPRS 63884. Arlington, Virginia: Joint Publications Research Service, 1975.

Kajoika, H., Yamaguchi, K., Sato, N., Soejima, K., and Sakaguchi, S. "Effects of Various Melting Parameters on the Qualities of Electroslag Remelted Ingots." In Proc. Fourth International Symposium on Electroslag Remelting Processes, pp. 102-14. Tokyo: The Iron and Steel Institute of Japan, 1973.

Kamenskii, L.A., et al. Sbornik Trud. Mosk. Verch. Met. Inst. 10(1971):146.

Kammel, R., and Winterhager, H. "Raffination von Metallen durch Elektroschlacke-Umschmelzen" (Refining of Metals by Electroslag Remelting). Zeit. fur Erzbergbau und Metallhuttenwesen 21(1968): 471-5.

Kelley, E.W. Manufacturing Process for Improved High-Strength Superalloy Sheet. Report No. AFML-TR-69-114. Wright-Patterson Air Force Base, Ohio: Air Force Materials Laboratory, 1969.

Kelley, T.N., Klein, H.J., Sun, R.C., and Venal, W.V. "Electroslag Remelting of Precipitation Strengthened Alloys." In Proc. Fifth International Symposium on Electroslag and Other Special Melting Technology. Pittsburgh: Mellon Institute, 1974.

Kelley, T.N., Junker, D.A., and Chen, R.C.H. "Comparison of Electroslag and Vacuum-Arc-Refined Superalloys." In Proc. Third International Symposium on Electroslag and Other Special Melting Technology, p. 125. Pittsburgh: Mellon Institute, 1971.

Klein, H.J. "The Feasibility of Producing Superalloy Electroslag Remelted Hollows." Journal of Vacuum Science Technology 9(Nov./Dec., 1972): 1334-8.

_____. "The Effect of a Variation of Melt Parameters on the Electroslag Remelting of a Nickel-Base Alloy." Proc. Electric Furnace Conference 28(1970): 13-9.

Klein, H.J., and Pridgeon, J.W. "Effective Electroslag Remelting of Superalloys." In Proc. Second International Conference on Superalloys Processing. B(1972).

Korenyuk, Yu. M., and Didkovskii, V.P. "The Electroslag Casting of Ingots of Copper and Some Copper Alloys." Automatic Welding 13(1960).

Kubisch, Ch., and Holzgruber, W. "Results Obtained with the Boehler Pressure ESR Process." In Proc. Third International Symposium on Electroslag and Other Special Melting Technology, Vol. III, pp. 267-84. Pittsburgh: Mellon Institute, 1971.

Kusamichi, H., and Fukuhara, Y. Tetsu-To-Hagane 52(1966): 1890-905. Translation No. 6542. London: The Iron and Steel Institute.

Lachner, W., Fiedler, H., and Muller, F. "Results with Plasma Torch Furnaces for Melting High Quality Steels from High Alloy Scrap." In Proc. Fourth International Symposium on Electroslag Remelting Processes. Tokyo: The Iron and Steel Institute of Japan, 1973.

Lakomsky, V.I. "Plasma Arc Remelting of Metals and Alloys." Summary Report, International Symposium on Special Electrometallurgy. Kiev: Russian Academy of Sciences, June 1972.

Latash, Yu. V., et al. Electroslag Remelting. Translation No. NTIS AD 73071. Springfield, Virginia: National Technical Information Service, 1971.

Latash, Yu. V., and Medovar, B.I. "Electroslag Melting." In Metallurgy. Moscow, 1970.

_____. "Electroslag Melting of Copper." Soviet Journal of Non-Ferrous Metals 7(1966): 84-5.

Lekarenko, E.M., Pokrovskaya, G.N., and Chernykh, K.P. "Electroslag Remelting of Monel Metal." Soviet Journal of Non-Ferrous Metals 5 (1964): 95-6.

Leybenzon, S.A., and Tregubenko, A.F. "Steel Production by Electroslag Remelting Method." Translation No. OTS:63-31866. Springfield, Virginia: National Technical Information Service, 1963.

Little, J.H., Queen, T.J., and Mackenzie, I.M. "Metallurgical Study of Electroslag-Refined Plates." In Electroslag Refining, pp. 43-53. London: The Iron and Steel Institute, 1973.

Lowe, E.M., and Hogg, A. "Application of ESR to Alloy-Steel Forgings." In Electroslag Refining, pp. 68-79. London: The Iron and Steel Institute, 1973.

Lowry, J.D., and Schlienger, M.P. "The Anatomy of a Non-Consumable Arc System." In Proc. Third International Symposium on Electroslag and Other Special Melting Technology. Pittsburgh: Mellon Institute, 1971.

Loyd, G.W., Owen, T.A., and Baker, L.A. Journal of the Australian Institute of Metallurgy 16(1971).

Luchok, J., and Wooding, P.J. Apparatus for Casting a Plurality of Ingots in a Consumable Electrode Furnace. U.S. Patent No. 3,834,447, CONSARC Corporation (September 10, 1974).

_____. Method of Casting a Plurality of Ingots in a Consumable Electrode Furnace. U.S. Patent No. 3,782,445, CONSARC Corp. (January 1, 1974).

Lukens Steel Co. Lukens Lectrefine® Steels. Technical Bulletin Form No. 994. Coatesville, Pennsylvania: Lukens Steel Company, 1973.

Machner, P. "Die Hauptparameter des ESU-Prozesses und deren Einfluss auf den Blockaufbau und die Wirtschaftlichkeit des Verfahrens." Berg-u Huettenmaennische Monatshefte 118(1973): 365-72.

Madano, O. "Production of Aluminum Iron-Base Alloys by Electroslag Melting." In Proc. Second International Symposium on Electroslag and Other Special Melting Technology. Pittsburgh: Mellon Institute, 1969.

McCullough, R.J. "A New Concept in Melting Metals." Journal of Metals (December 1962).

McKeen, W.A., Joseph, L.G., and Spehar, D.M. "Melting Alloys by the Hopkins Process." Metal Progress 82(1962): 86.

Medovar, B.I., Latash, Yu.V., Muksimovich, B.I., and Stupak, L.M. Electroslag Remelting. Translation No. JPRS 22217. Arlington, Virginia: Joint Publications Research Service, 1963.

Mitchell, A., University of British Columbia, Canada. Personal Communication, September, 1974.

Mitchell, A., Szekely, J., and Elliott, J.F. "The Use of Mathematical Models in Simulating ESR." In Electroslag Refining, p. 1. London: The Iron and Steel Institute, 1973.

Morozov, E.I., et al. "Electroslag Smelting of Titanium Ingots." In Titanium in Industry (Oborogiz, Moscow). Translation No. 16,114. Arlington, Virginia: National Technical Information Service, 1962.

Nafziger, R.H. "Slag Compositions for Titanium Electroslag Melting and Effects of Selected Melting Parameters." In Proc. Second International Symposium on Electroslag and Other Special Melting Technology, Vol. I. Pittsburgh: Mellon Institute, 1969.

Nafziger, R.H., and Calvert, E.D. "Electroslag Melting of Zirconium and Selected Properties of Fabricated Material." In Proc. Third International Symposium on Electroslag and Other Special Melting Technology, Vol. II, pp. 253-79. Pittsburgh: Mellon Institute, 1971.

Nafziger, R.H., and Lincoln, R.L. "Electroslag Remelting of Cobalt-Base Superalloys." Cobalt (1974): 79-85.

Nafziger, R.H., and Riazance, N. "Alkaline Earth Fluoride-LaF_3 Systems with Implications for Electroslag Melting." Journal of the American Ceramic Society 55(1972): 130-4.

National Aeronautics and Space Administration. Plasma Jet Technology. NASA Technology Survey Report SP-5033. Washington, D.C.: National Aeronautics and Space Administration, 1965.

Norcross, J.E. "Continuous Casting Heavy Wall Pressure Vessels Using the CESM (Continuous Electroslag Melting) Process." In Proc. Third International Symposium on Electroslag and Other Special Melting Technology, Vol. III. Pittsburgh: Mellon Institute, 1971.

Pateisky, G., Biele, H., and Fleisher, H.J. "The Reactions of Titanium and Silicon with Al_2O_3-CaO-CaF_2 Slags in the ESR Process. Journal of Vacuum Science and Technology 9 (November/December 1972): 1318-21.

Paton Electric Welding Institute. "Special Electrometallurgy." In Reports of the International Symposium on Special Electrometallurgy, Vol. I. Kiev: Russian Academy of Sciences, June 1972.

Paton, B. Ye., et al. "Mathematical Simulation and Prediction of Electroslag Refining of Large Forging Ingots. Electroslag Refining, p. 16. London: The Iron and Steel Institute, 1973.

_____. "Producing Super-Large Billets from Ingots or Forgings on the Basis of Electroslag Technology. In Proc. Fourth International Symposium on Electroslag Remelting Processes, pp. 376-82. Tokyo: The Iron and Steel Institute of Japan, 1973.

_____. "New Potentialities of Electroslag Shaped Castings." In Reports of the International Symposium on Special Electrometallurgy, Vol. I. Kiev: Russian Academy of Sciences, June 1972.

_____. "Electroslag Melting of Large Tonnage Ingots." In Proc. Second International Symposium on Electroslag and Other Special Melting Technology. Pittsburgh: Mellon Institute, 1969.

Paton, B. Ye., Lakomsky, V.I., Torkhov, G.F., and Philiptchuk, V.I. "Plasma-Arc Remelting in Copper Water-Cooled Crystallizer as a New Method of Improving Metal and Alloy Properties." In Proc. Third International Symposium on Electroslag and Other Special Melting Technology. Pittsburgh: Mellon Institute, 1971.

Paton, B. Ye., Medovar, B.I., Latash, Y.V., and Chekolito, L.V. "New Trends in Development of Electroslag Remelting." In Proc. Second International Symposium on Electroslag and Other Special Melting Technology. Pittsburgh: Mellon Institute, 1969.

Paton, B. Ye., Medovar, B.I., and Latash, Y.V. "Electroslag Remelting in the Soviet Union." In *Proc. First International Symposium on Electroslag Consumable Electrode Remelting and Casting Technology*, Vol. II. Pittsburgh: Mellon Institute, 1967.

Plasma-Arc Remelting of Metals and Alloys in Water-Cooled Copper Crystallizer. Licensintorg, Moscow.

Ploeckinger, I., Holzgruber, W., and Schneidhofer, A. *Electroslag Remelting Process and Apparatus for Producing Metal Ingots Having a Change in Transverse Dimension*. U.S. Patent No. 3,643,726, Boehler Company, Austria (February 22, 1972).

Pogodin-Alekseev, G.I., and Syrovatkin, A.A. "Electroslag Resmelting of Cathode Copper." *Soviet Journal of Non-Ferrous Metals* 5(1964): 82-4.

Pridgeon, J.W., Pochon, M.L., Gross, R.T., and Sharma, V. "Comparison of Hot Workability of Electroslag and Vacuum Arc Melted Nickel and Cobalt-Base Alloys." *Trans. International Vacuum Metallurgy Conference*, pp. 525-52. New York: American Vacuum Society, 1968.

Roberts, R.J. "Power Supplies for Electroslag Furnaces--A Preliminary Comparison." In *Proc. International Conference on Vacuum Metallurgy*, Anaheim, California. Columbus, Ohio: Battelle Memorial Institute, 1970.

_____. "Techniques for Maximum Power Consumption in Electroslag Remelting." In *Proc. Second International Symposium on Electroslag and Other Special Melting Technology*. Pittsburgh: Mellon Institute, 1969.

Savitsky, Ye. M., Burkhanor, G.S., Taskatov, N.N., and Shnyrev, G.D. "Production of Monocrystals of High-Melting Metals by Plasma-Arc Heating." In *Plasma Processes in Metallurgy*. Translation No. 61321. Arlington, Virginia: Joint Publications Research Service, 1974.

Schlatter, R. "Application of the Electroflux Remelting Process for High Speed Tool Steels." *Iron and Steel International* 47(1974): 197-204.

_____. "Electroflux Remelting of Tool Steels." *Metals Engineering Quarterly* 1(1972): 48-60.

Schmidt, F.A., Energy Research and Development Administration, Ames Laboratories, Iowa. Personal Communication, September, 1974.

Schneider, P.E. *Electric Furnace Proceedings* 27(1969): 189-97.

Schwerdtfeger, K., and Klein, K. "Rate of Fluorine Volatilization from Fluoride Containing Liquid Slags." In *Proc. Fourth International Symposium on Electroslag Remelting Processes*, pp. 81-90. Tokyo: The Iron and Steel Institute of Japan, 1973.

Stupak, L.M. "Electroslag Furnace ESR-40." In *Proc. Fifth International Symposium on Electroslag and Other Special Melting Technologies*, Vol. I, pp. 62-9. Pittsburgh: Mellon Institute, 1974.

Swift, R.A. "Electroslag Remelting Improves Properties of Plate Steels." *Metal Progress* (May 1973).

Swift, R.A., and Gulya, J.A. "Property Evaluation of Electroslag Remelted A533B Plate." *Welding Journal* 52 (December 1973): 537-s.

Thomas, A.G. "Direct Electroslag Melting (DESM) of Steel, Refractory Metal and Ferro-Alloys." In *Proc. Third International Symposium on Electroslag and Other Special Melting Technology*. Pittsburgh: Mellon Institute, 1971.

Ujiie, A., et al. "Production of High Temperature Alloy Tube by Newly Developed 'YOZO Technique'." In *Proc. Fourth International Symposium on Electroslag Remelting Processes*. Tokyo: The Iron and Steel Institute of Japan, 1973.

_____. "Application of Electroslag Remelting Process for Production of Heavy Thickness Material Pressure Vessel." In *Reports of the International Symposium on Special Electrometallurgy*, Vol. II. Kiev: Russian Academy of Sciences, June 1972.

Vachugov, G.A., Khlynov, V.V., and Khusim, G.A. *Stal.* 6, 483(1967).

Vainshtok, M.J., Moldavskii, O.D., and Kulakova, T.S. "Electroslag Remelting of Chromium Bronze." *Soviet Journal of Non-Ferrous Metals* 12(1971): 62-4.

Wahlster, M., and Schumann, R. "A Contribution to the Electroslag Remelting of Large Forging Ingots." In *Proc. Fourth International Symposium on Electroslag Remelting Processes*, pp. 337-45. Tokyo: The Iron and Steel Institute of Japan, 1973.

"Welding Equipment from the USSR." *Soviet Export*, Editorial Board.

Wilson, J.R., Hough, P., and Goble, M. "Electroslag Remelting of Aluminum-Magnesium Alloys." In *Proc. Second International Symposium on Electroslag and Other Special Melting Technology*, Vol. III. Pittsburgh: Mellon Institute, 1969.

Winterhager, H., Kammel, R., and Gad, A. *Ferschungsber, Nordrhlin-Westfalen* 2115 (1970).

Wooding, P.J., and Cerstvik, M.S. "Recent Developments in VAR and ESR and Their Impact on the Commercial Use of Consumable Electrode Melting." Paper No. 138 presented at Electro-Heat Conference in Brighton, England, May 1968.

Zeke, J., and Zelko, J. "Employment of the Electroslag Process of Melting for Welding of Very Heavy Sections and for Surfacing." In *Proc. Second International Symposium on Electroslag and Other Special Melting Technology*, Vol. II. Pittsburgh: Mellon Institute, 1969.

UNCLASSIFIED

SECURITY CLASSIFICATION OF THIS PAGE (When Data Entered)

REPORT DOCUMENTATION PAGE
READ INSTRUCTIONS
BEFORE COMPLETING FORM

1. REPORT NUMBER NMAB-324	2. GOVT ACCESSION NO.	3. RECIPIENT'S CATALOG NUMBER
4. TITLE (and Subtitle) Electroslag Remelting and Plasma Arc Melting	5. TYPE OF REPORT & PERIOD COVERED Final Report
	6. PERFORMING ORG. REPORT NUMBER NMAB-324
7. AUTHOR(s)	8. CONTRACT OR GRANT NUMBER(s) MDA-903-74-C-0167
9. PERFORMING ORGANIZATION NAME AND ADDRESS National Materials Advisory Board National Academy of Sciences 2101 Constitution Ave., N.W., Wash. D.C. 20418	10. PROGRAM ELEMENT, PROJECT, TASK AREA & WORK UNIT NUMBERS
11. CONTROLLING OFFICE NAME AND ADDRESS Department of Defense — ODDR&E Washington, D.C. 20301	12. REPORT DATE 216
	13. NUMBER OF PAGES 214
14. MONITORING AGENCY NAME & ADDRESS(if different from Controlling Office)	15. SECURITY CLASS. (of this report) UNCLASSIFIED
	15a. DECLASSIFICATION/DOWNGRADING SCHEDULE

16. DISTRIBUTION STATEMENT (of this Report)

This report has been approved for public release and sale; its distribution is unlimited.

17. DISTRIBUTION STATEMENT (of the abstract entered in Block 20, if different from Report)

18. SUPPLEMENTARY NOTES

19. KEY WORDS (Continue on reverse side if necessary and identify by block number)

Alloys	Ingots	Refractory Metals
Castings	Metallurgy	Specialty Steels
Electroslag Remelting	Plasma Arc Melting	Steels
Furnace Design	Recycling	

20. ABSTRACT (Continue on reverse side if necessary and identify by block number)

An assessment is made of the electroslag remelting (ESR) and plasma arc melting (PAM) technologies used in the United States in manufacturing a variety of materials and sizes of ingots. Significant metal quality improvements in surface condition, mechanical properties, solidification, structure cleanliness, and yield over air-melted material have accelerated ESR usage, particularly abroad. PAM has potential in recycling scrap, producing complex alloys, large monocrystals of refractory metals and their alloys, and independent control of heat input and metal feed. ESR has found significant applications in producing premium quality high alloy and specialty alloy steels, providing features such as desulfurization, retention of volatile alloying elements, and the capability to produce shaped ingots and castings. Many advantages and disadvantages are listed along with research and development to improve the technology and usefulness.

DD FORM 1473 EDITION OF 1 NOV 65 IS OBSOLETE

UNCLASSIFIED